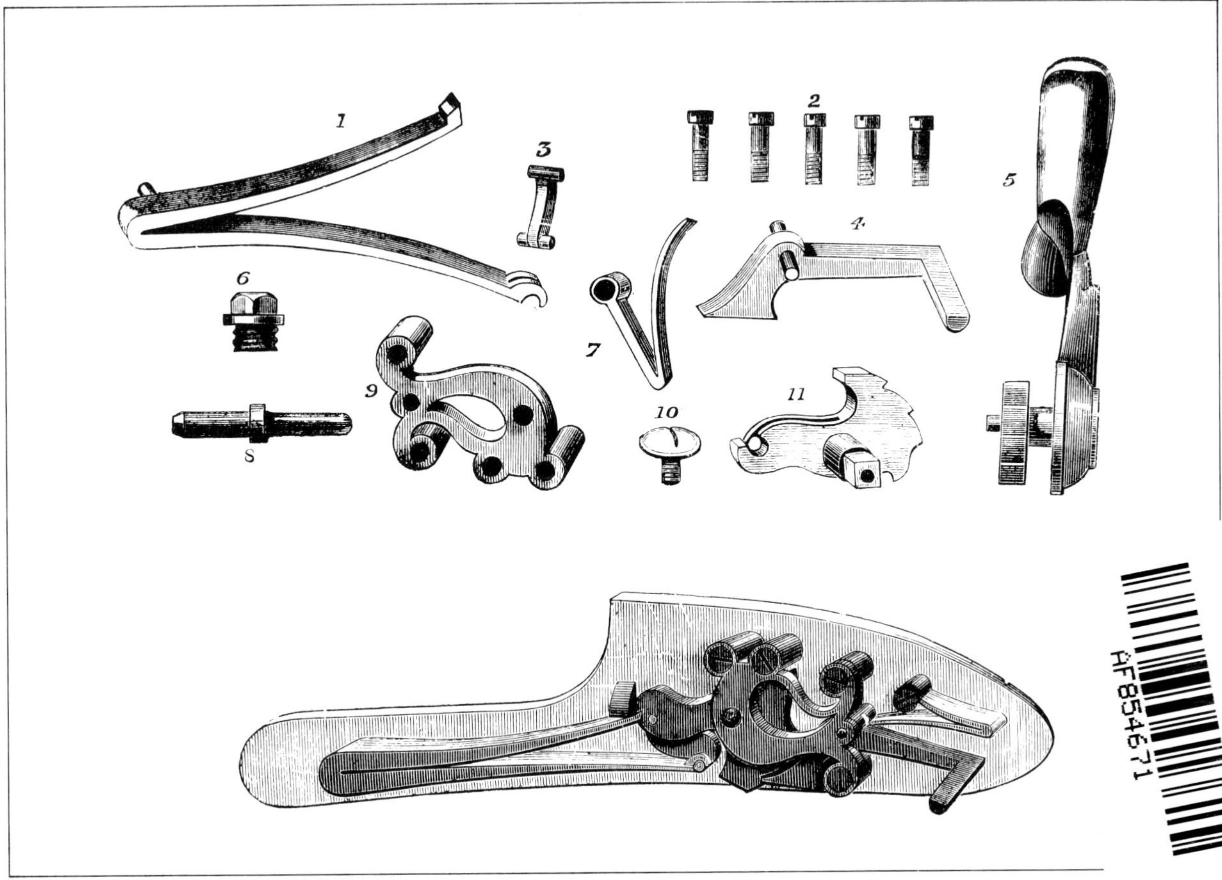
1
2
3
4
5
6
7
S
9
10
11

MODERN SHOT GUNS.

WOODCOCK-SHOOTING.

MODERN SHOT GUNS.

BY

W. W. GREENER,

AUTHOR OF "MODERN BREECH-LOADERS," "THE GUN AND ITS DEVELOPMENT," ETC.

Second Edition—Revised.

Tideline books

Sporting Book Specialists

RHYL. NORTH WALES. GREAT BRITAIN

1986

First published in 1888.
As a revised second edition
in 1891. This new edition
published by
Tideline Books
in 1986

Printed at The Bath Press, Avon

INTRODUCTION.

In the present treatise the Author has endeavoured to give such information as is constantly demanded of gunmakers by shooting men. Notwithstanding the number of published books written by sportsmen for sportsmen, there is much information indispensable to those who use guns, which the gunmaker only can impart. MODERN SHOT GUNS is not intended to be a sportsman's encyclopædia, it is not a compound of gunmakers' circulars, nor does it exhibit those conflicting opinions and diverse conclusions found in recent books on the same subject—such books having been written with the collaboration of various persons more or less interested in the subjects treated. These books do not enable the sportsman to arrive at any definite conclusion with respect to the merits or demerits of any mechanism, and merely "beat about the bush" with respect to the more important technicalities of gunnery. The columns devoted to shooting in the sporting newspapers teem with flippant nonsense concerning the facts of gunmaking, and although it is apparent that newspapers derive benefit by keeping subjects perpetually open, by so doing they propagate ignorance amongst their readers, and do but bewilder those they pretend to lead.

It is to the interest of those who use guns, as it is to the interest of the Author, that all the vexed questions of Gunnery be settled as far as is possible. There can be no excuse for the writers of books intended to guide sportsmen who fail to give a definite opinion with respect to the merits of "choke-bore" or "cylinder," of "hammer-guns" or "hammerless," and whether "nitro-compounds" are better or worse than black gunpowders for use in the modern Shot Gun.

In MODERN SHOT GUNS the Author has made a special endeavour to give the intending purchaser of a modern gun such directions as will enable him to obtain the article best suited for his purpose, to load it to best advantage, and to keep it in good condition ready for immediate use.

The sportsman, therefore, who wishes to know decisively with respect to any recent topic or controversy connected with the gun, has but to turn to the pages of this treatise.

The Author has thought it advisable to preface the book with a *short* historical *résumé* of the development of the Shot Gun, for seeing that the question of "chokes *v.* cylinders" has been seriously discussed so recently as 1887, it may be that the controversy of "breech-loaders *v.* muzzle-loaders" will be resuscitated, and it is possible that information concerning the merits of "firelocks" and percussion guns may yet be required. Meanwhile, the Author takes this opportunity to tender his thanks to the Editors of the Sporting Newspapers for the valuable services rendered to the gun trade generally by their energetic endeavours to obtain more perfect Shot Guns for English sportsmen, and for their able support to and kindly recognition of worthy inventions; and trusts that in the future, as in the past, *they* will direct their attention to the further development of the Shot Gun, rather than to futile discussions of long since accredited inventions.

In the arrangement of this treatise the Author has followed a method which appears to him the best suited to convey an accurate idea of the modern Shot Gun and its capabilities under varying conditions, without necessitating wearying and bewildering repetitions.

The HISTORICAL Section is necessarily meagre: those sportsmen who desire further information respecting the firearms of our forefathers will find fuller particulars in the Author's book—"*The Gun and its Development.*"

The DESCRIPTIVE Section contains short particulars of the leading types of modern Shot Guns. The numerous illustrations and absence of technical words from the text will be appreciated by those sportsmen who do NOT care for the jargon of the gun-shop nor the phraseology of science.

In the ANALYTICAL Section the various parts of the modern Shot Gun are detailed, with their technical names.

The part referring to the barrels will have interest for many, whilst the details concerning gun-stocks may be perused with benefit by even the most experienced sportsmen.

The TECHNICAL Section is a long one, and, unfortunately, teems with figures, resulting from an endeavour to include the results of all recent important experiments, which, although a knowledge of them is of little actual importance to the sportsman, could not be left unnoticed in a treatise professing to treat of Sporting Gunnery.

The SPECIFICAL Section is devoted to descriptions and particulars of such special guns as used for "pigeon shooting," "ducking," "wild-fowling," for "large game," shooting with ball; and treats, moreover, of small-bore guns, light short 12-bore guns, and contains tables specifying what any Shoulder Gun may be expected to average with any of the ordinary loads.

In the CRITICAL Section the Author has fearlessly and without prejudice criticised the various types of guns described in the work, or found in the gun-seller's stock. He has not abstained from giving his opinion, nor from quoting the opinions of others, on the important questions agitating the minds of sportsmen, and has pointed out to those about to buy guns how they may avoid being imposed upon by the ignorant or fraudulent vendors of spurious worthless weapons.

The PRACTICAL Section is just what it professes to be, viz., a series of practical instructions to the users of guns. This and the preceding section are the longest and most important in the treatise. The sportsman who peruses them, and follows the hints given, will not fail to save himself expense when purchasing arms and ammunition, but will be able to make the best use of his weapons, and shoot with confidence, ease, and pleasure.

W. W. GREENER.

Birmingham,
January 2nd, 1888.

NOTE TO SECOND EDITION.

THE first large edition of MODERN SHOT GUNS having been exhausted, and the demand still continuing, the Author has felt it incumbent upon him to produce this *revised reprint* of a book which, judging by its favourable reception, has been of service to many, and may prove of use to all who handle guns.

Birmingham,
January 1st, 1891.

CONTENTS.

PART I.—THE GUN.

SECTION I.—HISTORICAL.

SECTION II.—DESCRIPTIVE.

SECTION III.—ANALYTICAL.

PAGE

SECTION III.—CARTRIDGE CASES AND WADDING.

SECTION IV.—GUN CASES AND IMPLEMENTS.

PART III.—TRAP SHOOTING.

SECTION I.—LIVE BIRD SHOOTING.

SECTION II.—INANIMATE TARGETS.

APPENDIX.

MODERN SHOT GUNS.

PART I.—THE GUN.

SECTION I.—HISTORICAL.

EARLY GUNS.

THERE was much opposition made to the adoption of firearms as weapons of war by the nobles and knights of the fourteenth and fifteenth centuries. The firearm was not considered by them to be a knightly weapon for any purpose, and so guns were scouted by the sportsmen of mediæval days, who, if they were not themselves knights, were rarely in a position to disregard the dicta of the warriors.

The knights notwithstanding, the use of firearms became general for purposes of war, and the "poachers" of Germany seem to have been able to make use of them at an early period, for in the State papers of the Elector Augustus of Saxony, dated the 10th of October, 1555, is a prohibition against the possession of firearms by peasants and shepherds.

Small shot was known in Germany early in the sixteenth century, and in 1562, by a Government Regulation, its use was "entirely and absolutely" prohibited throughout the Duchy of Mecklenburg.

The weapons used for sporting purposes were not rifled, and were often loaded with two or more small balls; with the invention of the rifle early in the sixteenth century by the German gunmakers, a second class of firearm was created, which was rapidly perfected and quickly adopted for large game shooting, as well as for purposes of war. The Shot Gun was gradually evolved from the original hand firearm, but as a Shot Gun its history can only date from the introduction of double-barrelled arms.

The Sporting Shot Gun may therefore be termed a modern weapon. It was not until the middle of the seventeenth century that double guns were made sufficiently small or light to be used as sporting weapons. The development was very slow; the earlier guns had no ribs, and were furnished with wheel locks; in the eighteenth century ribs were added, and the flint lock and hammer employed. The gun then became essentially the sporting weapon; its clumsiness and slowness had hitherto prevented its general adoption, and the Flint-lock Gun was considered by the sportsmen in the first quarter of this century to have established itself firmly as an almost perfect sporting weapon. This gun, in its best form, was made by English gunmakers, and the particular style of Joseph Manton was considered superior to all others. His gun, in its latest form, is shown in the adjoining illustration. Such guns are now treasured

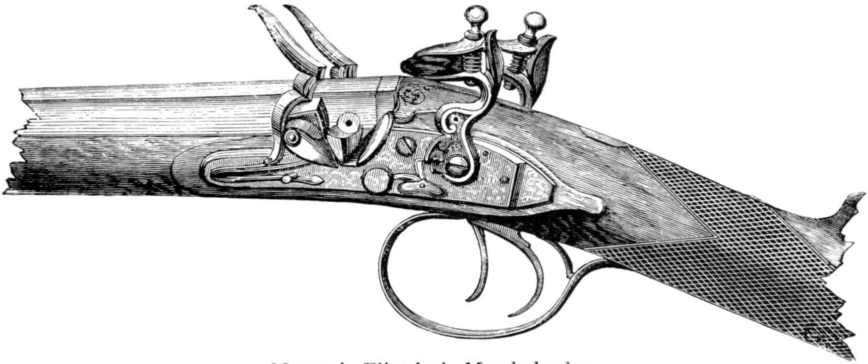

Manton's Flint-lock Muzzle-loader.

as curiosities by those fortunate enough to possess them, although the maker, and possibly their first purchaser, thought that fifty years would probably see them in practical use, not relegated to the limbo of forgotten lumber, or utilised to fill a museum gap or adorn the hall wall.

W. Greener's Double Muzzle-loader—1858.

Manton lived to see the introduction of the percussion cap, and a new era in the history of gunmaking commenced. Of percussion guns many forms appeared. All gunmakers, anxious to possess a speciality, invented detonating guns more or less practicable and useful, and of these many have been converted to the common form, and the others have joined the flint-locks in the region of the forgotten.

The percussion gun was a *great* improvement on the flint-lock, and although its day was short, it did much for English gunmakers, and if properly constructed was the most durable gun ever made, or probably ever to be made.

Fulminate seems to have been first applied successfully to firearms about 1806, and the name of the Rev. A. J. Forsyth, D.D., will always be associated with its practical application. From the date of his patent (1807) to 1825 many detonating guns were produced, and in 1818 the copper cap was made in England; it was not adopted by the English Government until 1840, although it had been generally used for sporting purposes some years previously. Before the percussion gun had arrived

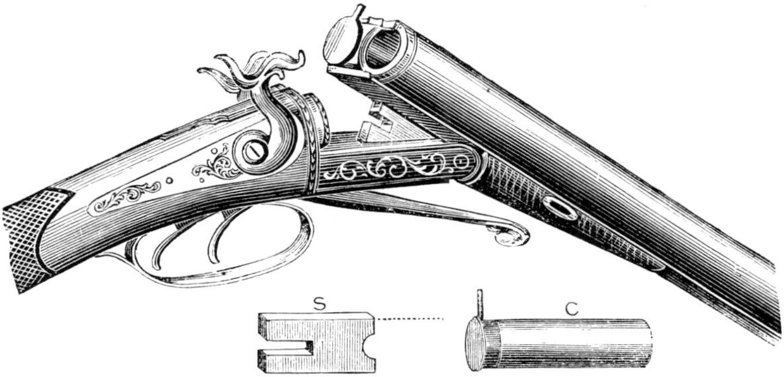

The Original Lefaucheux Breech-loader—1836.

at its highest state of perfection, the breech-loading shot-gun had made its appearance in France, and when improved by various English gunmakers bid successfully for public favour. The original Lefaucheux gun has long enjoyed, and, indeed, originally received a far greater share of praise than it actually merited. There was little really novel in the gun, and the mechanism, although simple, was ill adapted for the purpose. The Lefaucheux was favourably received, and the principle of the dropping down of the barrels for loading—which he adopted—has been faithfully adhered to since, and has now possibly attained its most perfect form, and for further decided improvements in modern sporting shot-guns the principle, not the details, will require alteration.

The pin-fire cartridge invented by Mr. Lefaucheux was a great achievement, and an important step towards the production of a perfect sporting gun. The shell or case, by expanding at the moment of discharge, effectually closes the breech-joint and prevents the escape of gas, and in this essential point is contained the secret of the success of the Breech-loading system.

The pin-fire cartridge was improved upon by English and French makers, and the central-fire gas-tight case, produced about 1853, marked a new epoch in the manufacture of sporting firearms. The first really successful central-fire shot-gun was the "Daw" gun, shown in the adjoining illustration ; it was enthusiastically championed by its introducer, Mr. Daw, of Threadneedle Street, and notwithstanding general opposition, its principle was soon adopted.

This gun was introduced about 1861, and became popular almost as soon as known. The central-fire system is still employed, and modern sporting guns are all made for the central-fire cartridge. It is the central-fire gun that has so changed the order of things. The percussion gun, as

Daw's Central-fire Hammer Gun.

an improvement on the flint, and the pin-fire, as an improvement on muzzle-loaders, are insignificant in comparison with the advantages of the central-fire system. Safety, speed, and ease are words which represent the main advantages, but they do not convey the full import to one unacquainted with the older types of sporting guns. Improvements during the first century and a half of the gun's adoption to sporting uses were scanty and apparently unimportant ; since the commencement of the present century each decade has seen numerous inventions applied to sporting guns, many of very considerable importance. There is no longer any need for writers to advance thirty reasons showing the superiority of the breech-loading over other systems, nor for sportsmen to fatigue themselves with heavy guns, and the annoyances of loading from the muzzle. The demand is for a gun that will do an immense amount of work, do it quickly, and with ease to the user. The large bags made almost daily during the season were a matter of impossibility to our grandfathers with their guns, providing even that other things were equal. The gun for the battue, for the field, and for the trap must fill requirements which Colonel Hawker or Squire Osbaldeston would have deemed preposterous, and certainly each decade, nay, each succeeding season, witnesses the production of new alterations to the sporting gun in such numbers that perfect as sportsmen may consider their guns, it is evident that to many disadvantages or imperfections are quite apparent.

The coming season will see further improvements, and many gunmakers will have novelties to the fore, but notwithstanding the fact that safe, reliable, hammerless guns have been more than ten years before the public, there are sportsmen who prefer the hammer gun; and although, personally, the author thinks that it should, as manufactured by several of our leading gunmakers, be relegated to the past, he will describe it fully and treat it as a modern gun, merely illustrating here and mentioning historically the Top-lever plain snap Hammer gun, which has many seasons enjoyed the confidence of sportsmen and the London trade.

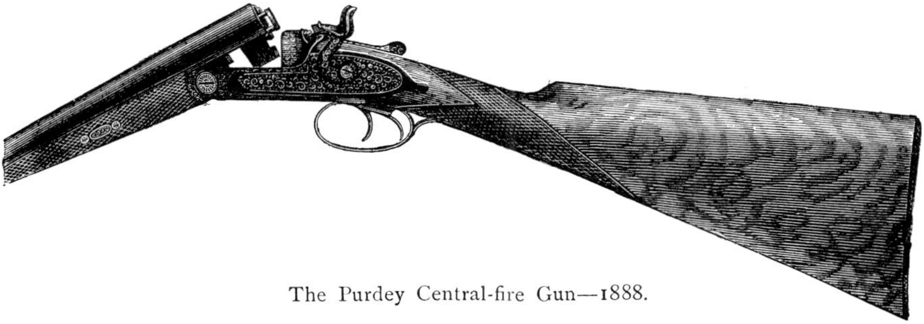

The Purdey Central-fire Gun—1888.

Many inventions, now so common as to be considered public property, and some of which are looked for in all modern guns, have in their time served to mark valuable improvements to the sporting shot-gun; first, to enumerate those of minor importance: the patent striker, the springless striker, the through lump, the circle joint, the spring fore-end, the "solid-headed" striker, the one-legged extractor, the double grip, and the "snap action" breech mechanisms, the extended rib, &c. Those of primary importance—the rebounding lock, the top bolt, choke-boring, the hammerless gun.

The rebounding lock, so simple and so efficient, was quickly adopted; it saved many accidents, although causing some. With hammer guns it is still used, and should be, but the rebounding lock and hammer gun is being so rapidly superseded by the various hammerless guns that the advantages, which were of such importance fifteen years ago, seem now little more than a name. The top bolt was the invention of the celebrated gunmaker, Mr. Westley Richards, whose name is so well and so honourably known throughout the world in connection with firearms. Naturally, his simple form was quickly improved upon; the advantages of the principle inculcated, others were not slow to copy or devise a more perfect method of uniting the barrels to the standing breech at the top, the point where in a drop-down gun the greatest strain is exerted. In 1873 the author produced his now well known "Treble-Wedge-Fast," or top cross bolt gun—a mechanism by which the barrels were secured to the breech action by a double holding down bolt and an independent bolt, of steel *binding* the barrels at the top to the *standing breech* itself, the whole forming a triple wedge of almost illimitable strength. This plan has stood the test of a decade, and is rapidly growing in popularity; on the expiration of the patent its use will be practically universal.

Of choke-boring too much cannot be said in praise. What the Breech-loader has done to increase the rapidity of firing, the Choke-bore has done to increase the range and efficiency of the weapon. Let it be taken for granted that the *raison d'être* of a gun is to shoot: this acknowledged, it is evident that the importance of the Choke outweighs all the improvements of the last twenty-five years. Many countries have laid claim to the invention of choke-boring; it is still a vexed question; and whether the honour belongs to the Americans, to Pape of Newcastle, to the French or to the Spanish gunmakers of mediæval times, all the author wishes to say is that, so far as he is concerned, the invention is his own. The form of Choke produced, and which has been generally accepted, and the method of producing it, are both of his invention, and for it he is not indebted to the Americans, to other gunmakers, nor to books ancient or modern.

Hammerless guns shall be treated upon fully in the descriptive chapters of this work. It will suffice to say here that they are the highest development of the Sporting Shot Gun, and are likely to grow in favour, and practically to supersede other forms of breech-loading guns now in use.

The author, in concluding this brief outline of the development of the Shot Gun, must remark that, although in his opinion the guns passed in review in this chapter are out of date, there are others who differ from him, and it must not be supposed for a moment that guns such as described cannot now be bought. Thousands of flint-lock muskets are made yearly, and *sold;* thousands of percussion muzzle-loaders are exported each year from England and Belgium; pin-fire guns are still manufactured, and so the other types which have once occupied the front ranks will find admirers; but all practical, intelligent sportsmen will not be thus contented, and such are invited to follow the author carefully in the descriptive and critical notes on the guns of *to-day.*

SECTION II.—DESCRIPTIVE.

HAMMER GUNS.

To sportsmen there is a seemingly endless variety of guns. All makers claim one or more specialities; and although the sportsman may know the type of gun he requires—for instance, hammer or hammerless—his choice is practically unlimited, to judge from the advertisement columns of the leading sporting papers. To give a description of every kind of gun would be but to reproduce a score or more of gunmakers' catalogues, and to leave the sportsman in no little confusion. Each gun now to be

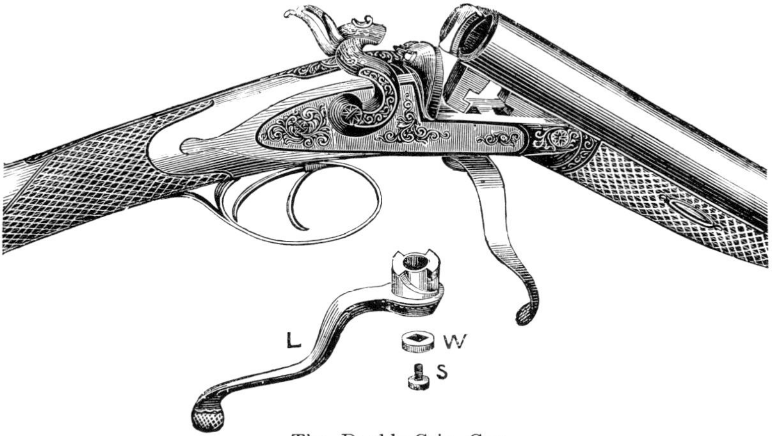

The Double-Grip Gun.

described must be considered as representing a type, the characteristics of which will be briefly, and the author hopes lucidly, set forth. The advantages or disadvantages will be more fully dealt with in a succeeding chapter.

The gun for several years the leading fashion, and even now not entirely out of date, known as the Double-Grip Gun, is the first, as it is also the oldest, pattern to be described.

Its appearance is probably well known to all. Its chief characteristic is the double-flanged lever-bolt—a modification of the original Lefaucheux

—working horizontally over the trigger-guard. There are no action springs; an inclined plane on the cylindrical head of the lever works against the barrel lump, and forces the barrels upward, when the lever is turned from the trigger-guard. When the gun has been closed, the lever is returned, and two flanges on the cylindrical head enter into notches in the barrel lump, drawing the barrels firmly down to the breech-action body. The mechanism is exceedingly simple; all the parts are strong, and, with back-action locks, it is a form of breech-action which, but for the time it takes to manipulate it, would be of great service for guns to shoot large charges of powder, and for hard work generally. It is now usually applied to low-priced guns only, and is rarely found upon very light guns. A modified action, in which the lever is worked by a spring, has been made.

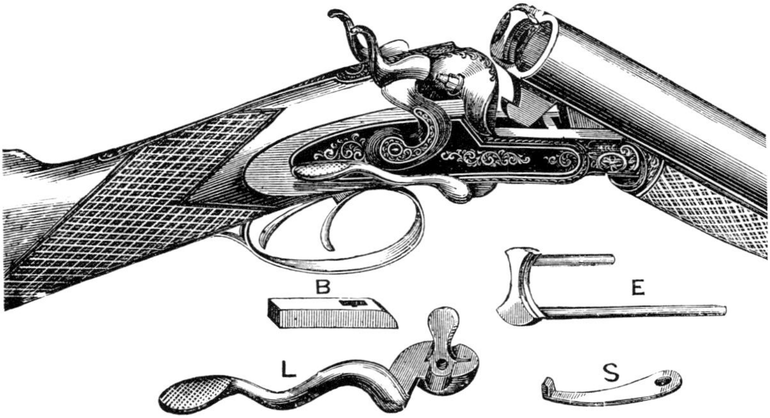

The Side-Lever Gun.

Another style of breech mechanism is the side lever. In this gun a lever is bent round from underneath the breech-action body to lie on the lock plate, its thumb-piece being conveniently placed immediately behind the hammer. This lever is pivoted beneath the barrels in the breech-action body, an arm being continued upwards to engage in a slot in a steel bolt, working longitudinally in a slot in the breech-action body—being forced forward by a spring, and moved backward by the lever. The bolt enters one or more notches in the lump underneath the barrels, and serves to *hold down* the barrels to the breech-action body.

The spring—a flat one—is sometimes fixed in the breech-action body, forward of the lever, and is uncovered; but it should be placed on the trigger-plate, and connected to the lever by an S swivel, similar to that of a gun-lock. Such guns work much more pleasantly, and the spring is not so likely to fail.

The side lever may be on either the right or left side of the gun.

A modification of this breech mechanism is the under lever, in which the lever, instead of being taken to the side, is bent to the form of the trigger-guard, and lies over it (see Daw's breech-loader, already illustrated). The lever is depressed to open the gun. When the gun is closed, the holding-down bolts snap home, carrying the lever to its place.

This mechanism is now occasionally supplemented by a bolt or bolts, engaging with an extension of the top rib of the barrels; but, although once much sought after, the side lever has been discarded by most gun-makers, and, with one notable exception, is to be found only in low-priced guns.

THE TOP-LEVER GUN.

The Top-Lever Gun is of two kinds: in one, the lever swings upon a horizontal pivot, in the tang of break-off or in the standing breech, and is

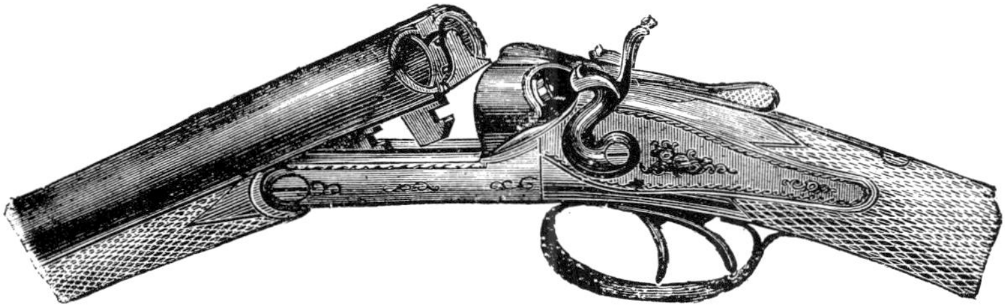

Top-Lever Gun, with Back-action Locks.

raised up to open the gun. This type is not generally used, and is seldom found, except upon the guns of two Birmingham makers.

The other type of lever is generally used; the lever turns upon a vertical pivot, and is pushed to the right to open the gun. An arm upon the pivot is connected with the action-bolt, and draws back the bolt to allow of the gun being opened, whilst a spring, connected with the bolt, the lever, or the pivot, forces the bolt into the bites on the barrel-lump as soon as the barrels are in their place for firing. The general appearance of a gun with this lever may be seen by referring to illustrations of the Purdey Gun, in the first chapter. Various mechanisms are used by different makers to bring about the withdrawal of the bolt or bolts; but there is, in reality, little to choose between the methods employed. The adjoined illustration shows a Top-lever hammer gun, with back-action locks; and it may as well be remarked here, that in a light gun the back-action lock allows of more metal, and consequently greater strength in the breech-action, whilst the grip of the stock is not proportionately weakened. In a heavy gun, the strain upon the hand of the stock is greater, and as the back-action lock does take from the strength of the stock in its weakest place, bar-locks should always be preferred upon heavy guns. Back-action lock guns are usually a little cheaper than bar-guns, although the difference is so small, if everything be of the best quality, that no alteration in the price is noticeable; but in a very cheap gun, where every shilling is of importance, the back-action lock is often used.

THE DOLL'S-HEAD GUN.

In this breech mechanism, the barrels have an extension of the top-rib,—or a separate steel lump equal thereto—which extension is let into a correspondingly shaped hole in the top of the standing breech.

The belief is that this head keeps the standing breech from springing

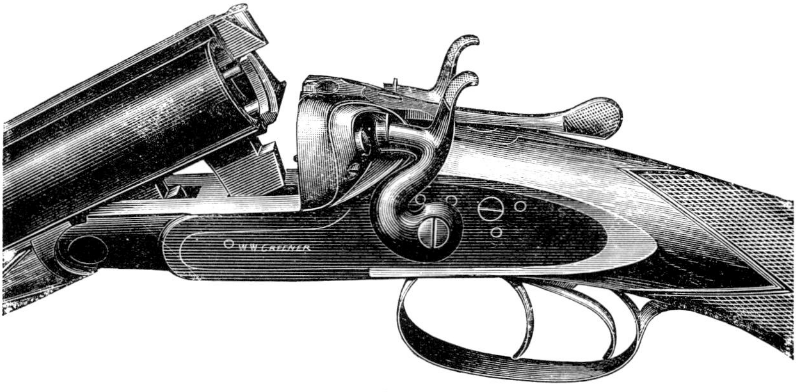

The Doll's-Head Gun.

back at the moment of discharge, and consequently increases the solidity of the weapon. For proof that this is not generally the case, the reader should refer to the Critical Section of this work. The only objection to the Doll's-Head Gun is that the somewhat large projection is in the way when loading, or when taking out the fired cases. The projection should always be used as a guide for the extractor; a groove in it will keep the extractor in its place better than the second leg usually employed.

The Doll's-Head is decidedly popular. In its primitive form, it is not generally found upon high-class weapons, but on guns of the third and lower grades it is considered to be a necessity.

The price of a gun as illustrated will average thirteen guineas in England, or 75 dollars abroad. Many very much cheaper guns are in the market, and the Doll's-Head will be found upon all the cheap grades in the American markets.

TREBLE-GRIP GUNS.

When, in addition to the usual double holding-down bolt, a gun is furnished with a bolt, engaging with the extension of the top rib, it is called a Treble-Grip Gun. The ordinary Doll's-Head Gun is sometimes so styled, but wrongly so. The crude idea of the ordinary Treble-Grip Gun would seem to have originated from a combination of the well-known Westley Richards' top-grip breech-action with the double holding-down bolt; but, strange to say, this is almost the last form the Treble-Grip Gun has taken. The well-known and very much superior Treble Wedge-Fast Gun—to be described—preceded it, as did many others of considerable worth. The third grip may be a prolongation of the top lever, a small

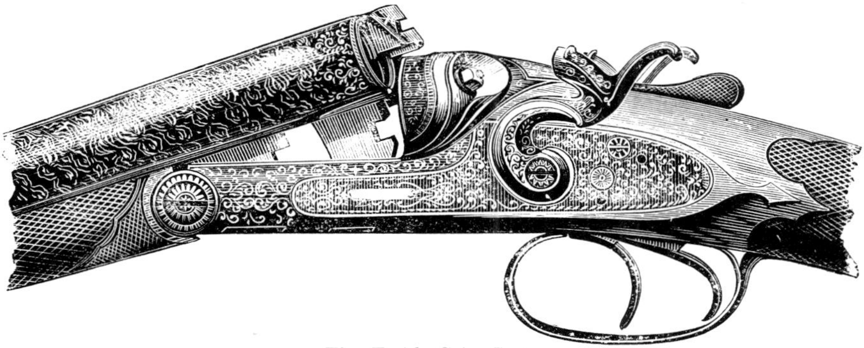

The Treble-Grip Gun.

bolt actuated by it, or a fancifully-shaped and named head engaging with slots or V grooves in the projecting rib. The Treble-Grip Gun illustrated is one of the simplest and best of the many forms now common. The third grip is a plain bearing of a prolongation of the top lever upon the projecting rib, and it not only materially lessens the strain upon the under bolts, but also keeps the projecting rib, which is dovetailed into the standing breech, up to its work. Many of the "Triplex Grip," "Climax," "Hold-fasts," "Giant Grip" guns have not half the wear in them this gun has, and the mechanism of such-named guns should be most carefully examined before purchase.

This style of gun may be purchased from 15 to 40 guineas, and most guns of second quality are furnished with a third bite, equal to one of the Treble-Grips already enumerated.

THE TREBLE WEDGE-FAST GUN.

This is decidedly the most popular English gun; it has numerous imitations, and it may therefore be said to have genuine merit. As this chapter is mainly descriptive, the reader should refer to the succeeding sections for critical remarks upon the various breech-loaders described here.

The Treble Wedge-Fast Gun was introduced by W. W. Greener in 1873, who invented it as an improvement upon the many flimsy snap-action breech-loaders which then had a certain popularity. It consists of a steel projection from the top rib, which fits into a slot in the standing breech. A round steel bolt, actuated by an arm of the top lever, works transversely in the standing breech and passes through the steel projection, binding the top of the barrels securely to the breech, so that any gaping or wear at the joint is impossible. Nothing more simple nor so efficient can be imagined. This top bolt is in itself fully equal to the strain of firing heavy charges, but in connection with the double holding-down bolt it

W. W. Greener's Patent Treble Wedge-Fast Gun.

works smoothly, and forms the strongest combination known, which is applicable to guns upon the drop-down principle.

The mechanism, so far from adding to the weight of the gun, diminishes it, for guns may be made on this principle lighter and stronger than if made upon any other. In appearance the Treble Wedge-Fast Gun is graceful; one can see at a moment that the chief quality is solidity, but it is impossible to find a trait of weightiness.

The Top-lever, from its position, far in arrear of the standing breech, is most handy, and the narrow projecting rib in no way interferes with the rapidity with which the cartridges may be inserted or withdrawn.

The mechanism is equally applicable to front and back-action locks, and is made on both plans, and is also applied to various hammerless guns. This gun is more expensive to produce than Treble-grip guns, and if well made, it is certainly without any equal for strength or beauty.

This is the gun for all who use heavy charges, for all who work their guns hard, for all who require a light and safe gun. The price may be said roughly to range from fifteen to fifty pounds; but on the Continent wretchedly made imitations are circulated, which have only the barest outward resemblance to the original English gun, which is the sole property of the patentee and inventor, W. W. Greener.

The Treble Wedge-Fast Gun has so much to recommend it that for almost any sport or purpose it may be chosen, and the greatest confidence may with safety be reposed in it, for the heaviest charge of any explosive is as powerless to injure it as are the sands of the Sahara or the snows of Siberia to stop its proper working.

SEMI-HAMMERLESS GUNS.

Self-cocking and self-half-cocking guns, with hammers, enjoyed a certain popularity in the 1860-70 decade. With pin-fire guns, the raising of the hammer to half-cock by means of the action lever was an advantage.

Central-fire guns, with rebounding locks, quite superseded all guns half-cocked by the lever. Hammer guns cocked by means of the action lever

The Semi-Hammerless Gun.

have every disadvantage of hammerless guns, with but one of the corresponding advantages. An attempt was made about 1876, by some gun-makers, to substitute a self-cocking gun for the hammerless, but the attempt signally failed.

Another compromise is the semi-hammerless gun, in which fingerless hammers are placed upon the outside of the lock plate, instead of being arranged within the gun.

The advantages claimed are that the gun is neater in appearance: that there can be no mistake as to whether or not the gun be at full cock. There is, however, nothing to show that the gun is *loaded;* all guns should always be treated as loaded; accidents would then be rare.

The Semi-hammerless cannot be said to be safer than the hammerless when all things are taken into consideration; it is somewhat more complicated, and is not likely to find favour with the generality of sportsmen.

EARLY HAMMERLESS GUNS.

A hammerless gun on the drop-down principle, the lock mechanism similar to the Prussian Needle-gun, has been in use in Germany for many years. Mr. Daw introduced a hammerless central-fire gun about 1862, but it did not attain the popularity of his central-fire hammer gun, and was practically abandoned. In 1866 a patent was taken out for a hammerless gun, in which the lock mechanism was attached to the trigger-plate, and the cocking effected by the under action lever.

The Murcott Hammerless Gun.

Mr. T. Murcott's hammerless gun, patented in 1871, is typical of the first hammerless and the present cheap hammerless guns.

The one drawback to this gun is the position of the lever. An under lever is neither so handy, so pleasant, nor so quick as a top lever. In the original Murcott the lever had to travel a long distance and required considerable force—faults which in a modern gun are remedied to a very great extent. The Gibbs and Pitt is a modification of the Murcott plan; Woodward's Automatic, Reeves', Lang's, and several others, are variously improved forms of the original idea, and the Murcott Gun being now manufactured cheaply, has become very popular with gun-dealers, and in one of its many forms may be found in almost every gun-shop. The price for a sound gun on this principle will average £20; some cheap modifications are to be found as low as £10, but are not to be recommended, but any one who, having a preference for the under lever, wishes a hammerless gun at £25, or thereabouts, should obtain a sound weapon on this principle.

A very simple, effective, and ingenious safety bolt was applied to all

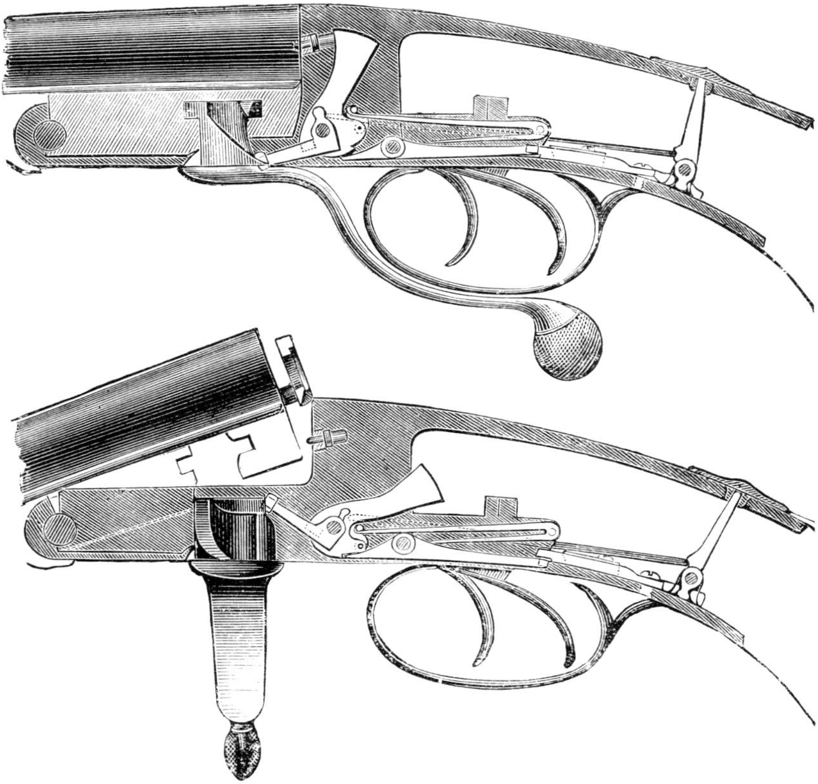

Allport's Double-grip Hammerless Gun.

the early Murcott Hammerless Guns; its position upon the top of the standing breech was not a handy one, but the bolt itself was very much better than the majority of complicated mechanisms misnamed "safeties," which are applied to the modern modifications of the weapon.

ALLPORT'S HAMMERLESS GUN.

This gun consists of an ingenious cocking mechanism and simple lock, combined with the well-known, and at one time extremely popular, double-grip breech-action.

The mechanism will be easily understood after reference to the illustration.

This gun works without any appreciable strain, is easy to cock, and being made only in the best style, may be strongly recommended to those sportsmen who prefer the double-grip breech-action.

A simple trigger-bolting top safety, in connection with an ingenious mechanism for locking the scears, is usually employed with this gun, but various safety bolts may be used.

The mechanism is adapted for double and single rifles and guns, and from its strength, simplicity, and durability, will doubtless receive the success it merits. A good ordinary 12-bore shot gun on this principle should in the shop of a London gun-dealer cost from £30 to £45; the

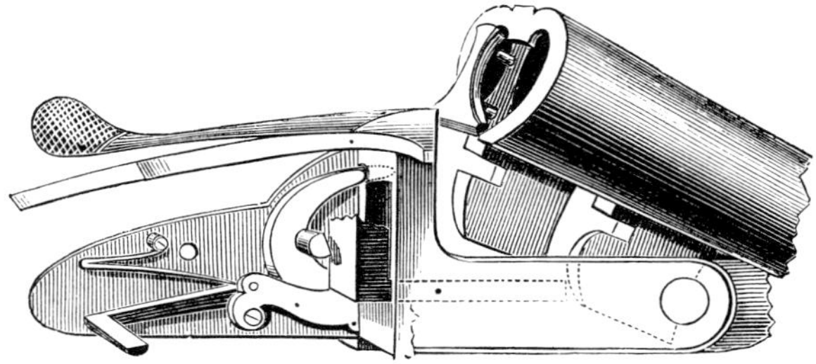

The "Club" Hammerless Gun.

gun is not made in very cheap grades. There are other double-grip hammerless guns besides the "Allport," which is taken as typical of the class.

TOP LEVER HAMMERLESS GUNS.

There are several patent hammerless guns with top levers, in which he cocking of the locks is effected by connecting the breech-action lever with the lock mechanism, so that upon opening the gun for loading the tumblers are forced into full cock. The top lever is very popular, but there are many difficulties to be surmounted before a hammerless gun in which the cocking is done by the lever, can be produced. The top lever is short, and the leverage obtained barely sufficient to easily withdraw the action bolts, and compress the action spring. The travel is short, although sufficient for withdrawing the action bolt; if the lever be made to travel further it is not handy, so although many arrangements of levers, friction rollers, and inclined planes have been tried, it is not surprising that a really good gun on this plan has not yet been produced.

By causing the holding-down bolt to engage with the barrel lug at the me the barrels are dropped for loading, and utilising the weight of the

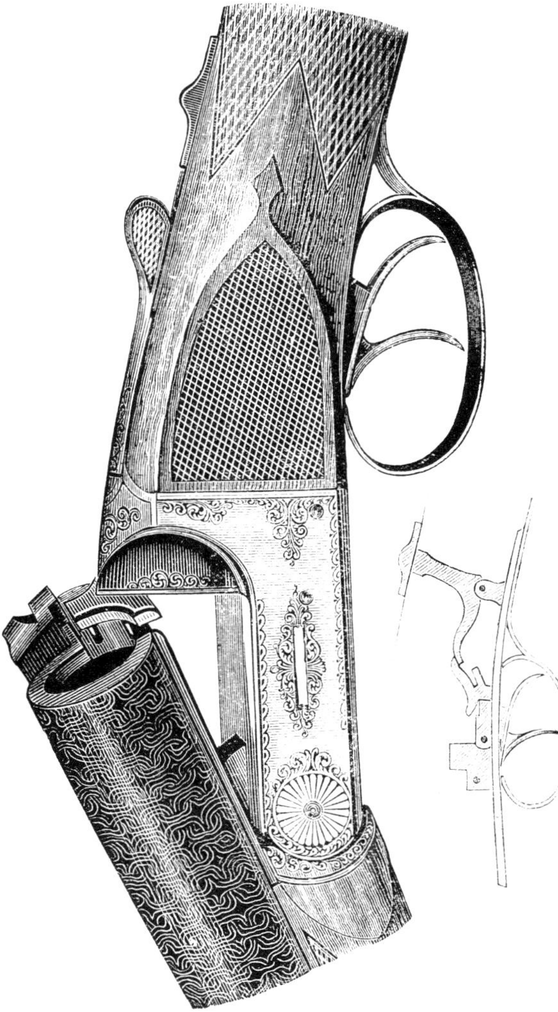

The Westley Richards (Anson and Deeley Patent) Hammerless Gun.

falling barrels as well as the action lever, it has been found possible to cock the locks without undue exertion. Several guns were once popular which were made upon this principle, but although they are not now so general, several gunmakers still offer them.

The "Club" Hammerless Gun made by W. W. Greener is typical of this class of gun. In London such guns are offered at £40, but they are also made in plain style for export and the country trade, and may be met with at half this price. Several safety bolts are applicable, and one or more will be generally found on all such guns.

THE ANSON AND DEELEY HAMMERLESS GUN.

This gun was produced in 1875, and for several years enjoyed a popularity never before attained by a hammerless gun. The cocking is effected by levers pivoted on a common centre with the barrels. The lock mechanism, which is particularly simple, is arranged within the breech-action body. The cocking levers engage with arms of the tumbler extended forward, and when the barrels are dropped for loading, the cocking levers move with them, and carry the locks into full bent. The joint pin is solid. The gun about the breech has a very square box-like appearance. Several systems of holding down bolts are used in conjunction with the Anson and Deeley lock, but the top-lever and a treble grip is the most general.

The safety usually employed is fixed upon the top, and cuts right through the weakest part of the gun-stock. The gun is made in many qualities, and may be found with the name of almost every maker—English and foreign. It has been popular in the United States under the name of the "Harrington and Richardson" Hammerless.

The cheapest guns made on this plan are made with loose hinge pins, sham top fastenings, and sometimes without any top connection whatever—all such guns should be avoided. They are offered at prices ranging from £10 to £20.

A serviceable gun on this principle, made by a reliable manufacturer, should be purchased for £15, and of course better quality guns. The best qualities of some makers are made upon this principle.

By a later patent the lock mechanism is modified to permit of the gun being put together without first cocking the locks; it is not so popular as the original form, and is somewhat dearer.

SCOTT'S HAMMERLESS GUN.

This hammerless gun was introduced to the gun trade in 1878, and being made at wholesale for retailers, quickly became popular. As it is now offered under many different names, it will be necessary to illustrate the mechanism in order that it may be readily distinguished. The cocking is effected by the dropping of the barrels for loading; the mechanism consists of rods (A) moving diagonally in the breech-action body, which rods have a notch near their fore extremity, with which studs, C, fixed under the barrels engage, drawing forward the rods as the breech-ends of the barrels rise. The rear extremity of each rod, A, engaging at B with the tumbler of the lock, pulls it into cock as the rod travels forward. The lock mechanism is affixed to side lock plates, which are usually furnished with crystal apertures, H, through which the position of the tumbler—which is often gilt—may be seen.

Several improvements have been added to this gun, and it may now be opened and closed with ease, and is pleasant to use.

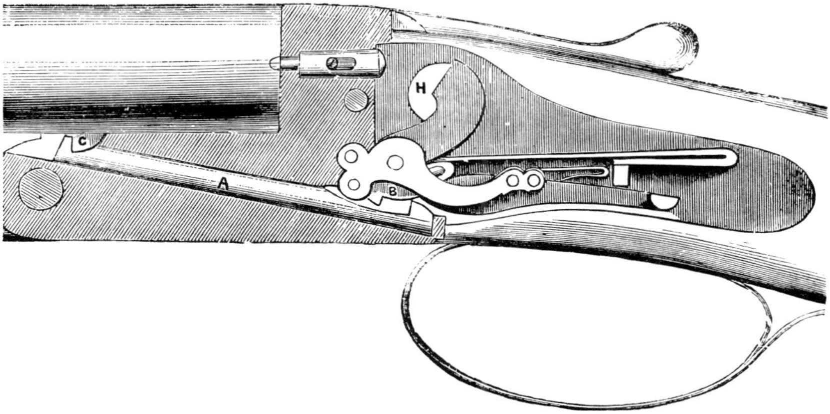

Mechanism of Scott's Hammerless Gun.

The chief points by which it may be differentiated are—the crystal apertures, H, the studs, C, under the barrels and gas-checks. These checks consist of blued-steel discs let into the face of the standing breech, having the strikers for their centre. A groove is made near the circumference of the discs, which joins a similar duct running horizontally across the standing breech to each extremity. This check may be likened to the trench dug round a bell tent, and its purpose is to provide an exit for any gas that may blow from the cartridge by reason of a badly-fitting cap or faulty cap dome.

The Scott Gun, under one or other of its many aliases, may be purchased at prices from 18 guineas to more than double that amount.

PURDEY'S HAMMERLESS GUN.

Another principle for the lock and cocking mechanism of hammerless guns is illustrated in the Purdey Hammerless, which is typical of several patented systems possessed by London gunmakers of considerable renown

Here, again, the mechanism must be illustrated to convey any adequate idea of its working. The principle is that of a rebounding lock, which, instead of rebounding to half, goes to full cock immediately the gun is

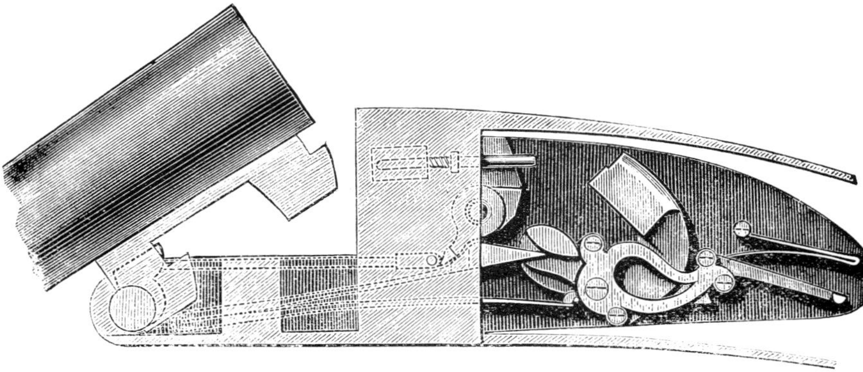

Purdey's Hammerless Gun.

opened. The spring is cramped or set as the gun is closed, by means of the rod and cam shown in dotted lines in the accompanying illustration.

Sporting guns made on this principle may be neatly and symmetrically

W. W. Greener's Treble Wedge-Fast Hammerless Gun.

fashioned, and have the appearance of compactness which the ordinary front action lock conveys. It is somewhat hard to close, but has fewer drawbacks than many hammerless guns, cocking with the opening of the barrels. The price of this gun is about seventy guineas.

THE "GREENER" HAMMERLESS GUN.

In this gun the cocking is effected by the barrels. In general appearance the gun has not the squareness or box-like form about the breech-action common to the Anson and Deeley, and several other hammerless guns. The breech-action is neat, rounded beneath near the fore end, and not greatly differing from an ordinary gun in size.

The substitution of a screw hinge-pin for the solid pivot of the Anson and Deeley adds to the strength and solidity of the breech-action.

In the lock mechanism the tumblers are roughly of the nature employed

Mechanism of Greener's Hammerless Gun.

by Needham and Anson and Deeley, but are acted upon by a swivel catch, pivoted in the under lump of the barrels. The above illustration shows the arrangement of the mechanism. As the barrels are dropped for loading, the breech-ends rise, and the swivel, engaging with the turned-in extremities of the tumblers, carries them to full cock, the weight of the barrels being sufficient to carry both tumblers to full bent. There are no "lifting dogs." The gun is readily taken apart and put together, it being immaterial whether the gun is cocked or not upon replacing the barrels. The breech mechanism is the "Treble Wedge-fast" already described, and this permits of a stronger and lighter hammerless gun being made than has yet been constructed upon any other principle. This gun is readily obtainable at prices from sixteen to fifty guineas.

HAMMERLESS "EJECTING GUNS."

Modern sporting breech-loaders, like perfected military firearms, eject the fired case. This principle is certain to obtain popularity. There are difficulties in the way; but it should be remembered that every cartridge case after being fired contracts, and consequently is or should be comparatively loose in the chamber, requiring but little force of the right kind to effect its dislodgement.

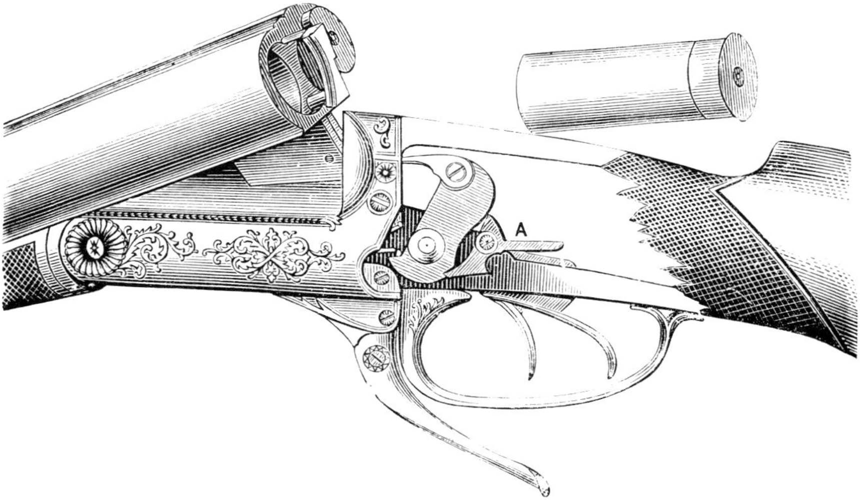

THE NEEDHAM EJECTOR GUN.

With Mr. Joseph Needham originated the idea of having a separate extractor for each barrel for the purpose of using self-acting mechanism to extract the cartridge case.

In the Needham Ejector Gun a separate extractor is used for each lock; the tumbler of each lock when released from bent—*i.e.*, when the gun is fired—falls into such a position that the extractor mechanism will engage with it as the gun is opened, and the fired cartridge be thrown clear of the gun. The ejecting force lies in the mainspring of the lock, and is in no way dependent upon the lever, nor does it require a special effort from the shooter.

In the Needham Ejector the barrels are forced upwards by the lever; a modification is also made with a top lever, but even in its cheapest form the gun is well worth the thirty or more pounds asked for it.

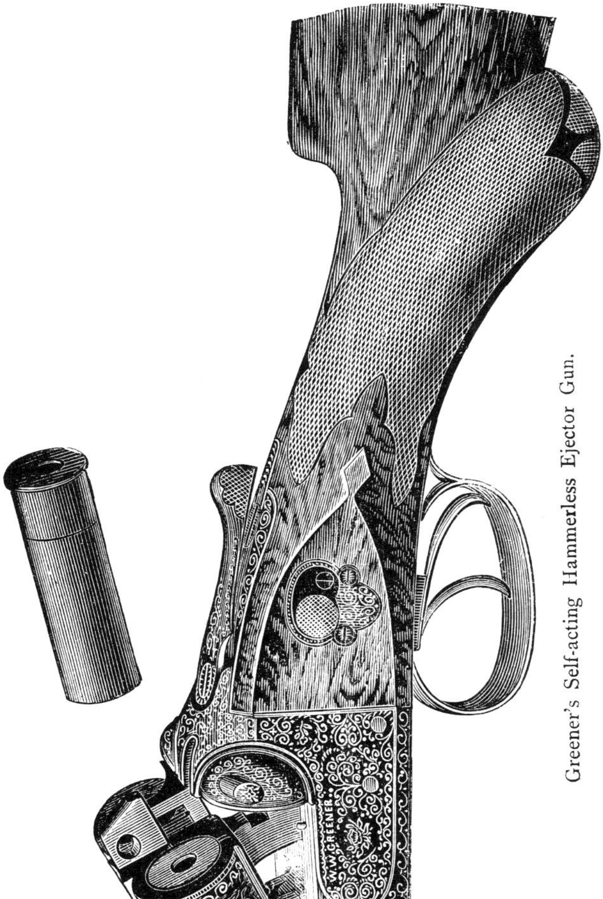

Greener's Self-acting Hammerless Ejector Gun.

GREENER'S SELF-ACTING EJECTOR GUN.

This gun possesses a breech-action that is self-fastening, a lock that is self-cocking, an extractor that is self-ejecting. The mechanism is simple, and in principle akin to that of the Needham Ejector Gun. The breech mechanism is the patent Treble Wedge-fast; the lock mechanism that of W. W. Greener's Hammerless Gun already described. The cocking mechanism is the same, modified to admit of the ejecting mechanism.

This modification consists of the addition of a stud, which is left in the cocking swivel about midway: immediately beneath this stud project the lower extremities of two ejecting levers, pivoted in the barrel lump and

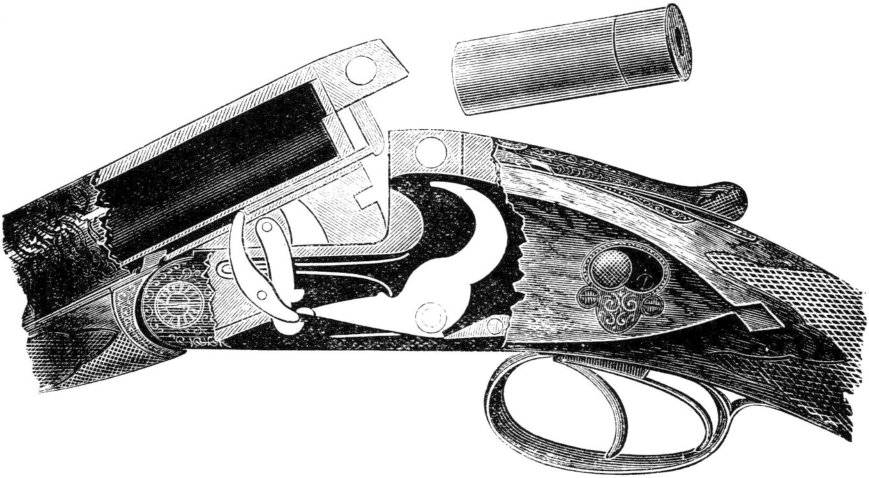

Mechanism of Greener's Ejector Gun.

communicating with the legs of the extractors. Each extractor acts independently of the other, and in connection only with its own barrel and lock, and thus it is impossible for an unfired cartridge to be thrown out.

Presuming that the gun has been fired, the action is as follows:—On opening the barrels, the tumblers are raised by their turned-in forward extremities bearing on the additional stud of the cocking swivel. When nearly to cock, they slip past the stud and fall sharply upon the ejector's lower arms, and the extractors, already forced partly out by a lever in the fore-end in the usual manner, are violently propelled to their full extent by the blow, and flip out the fired cases.

If one cartridge only be fired, the other lock remaining at cock does not engage with the cocking swivel or ejecting lever; consequently unfired cartridges are simply withdrawn to the ordinary extent.

The power available for ejecting the cases is that of the mainspring falling, from reasons already explained; the ejecting of the fired case is effected perfectly.

The gun requires most careful adjustment, and—although the parts are few and most simple—to ensure perfect working the utmost precision is necessary in centreing and shaping the various limbs. Consequently, the gun can only be made by experienced workmen, and must be made of best quality throughout. The prices, therefore, range from forty guineas for a gun plainly finished to sixty for those elaborately ornamented. The principle of ejecting may be used with suitable mechanism upon various hammerless and hammer guns, but owing to the success of the author's ejecting gun, somewhat similar, but less perfect, mechanisms are in the market, whilst there are numerous imitations so very inferior that the sportsman will readily shun them.

THE DEELEY EJECTOR GUN.

This ejector gun is of a different type to the Needham, extra lock mechanism is provided for the purpose of expelling the empty cases; and

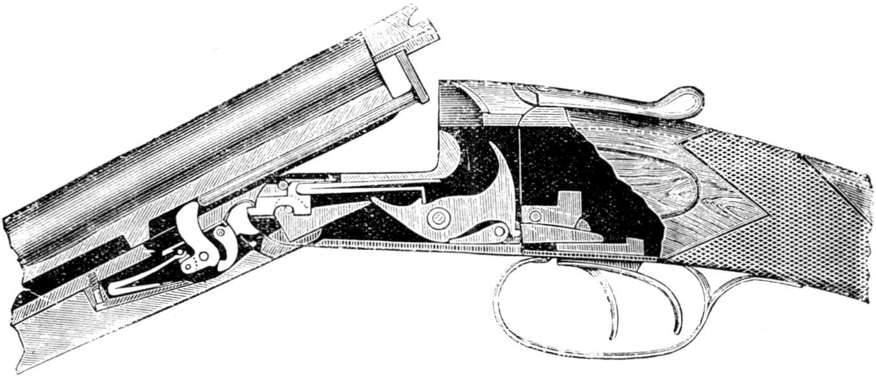

The Deeley Ejector Gun.

in the Deeley gun this additional mechanism is placed in the fore-end of the gun. The hammer of the second lock, when released by its scear, strikes the leg of the extractor and flips out the fired case; the scear is liberated by the falling of the striker of the ordinary lock altering the position of its mainspring, pushing it forward nearly one-eighth of an inch towards the muzzle of the gun, and thereby coming into contact with the mechanism provided for the release of the ejecting hammer.

The ordinary lock mechanism is that of the Anson and Deeley gun, and

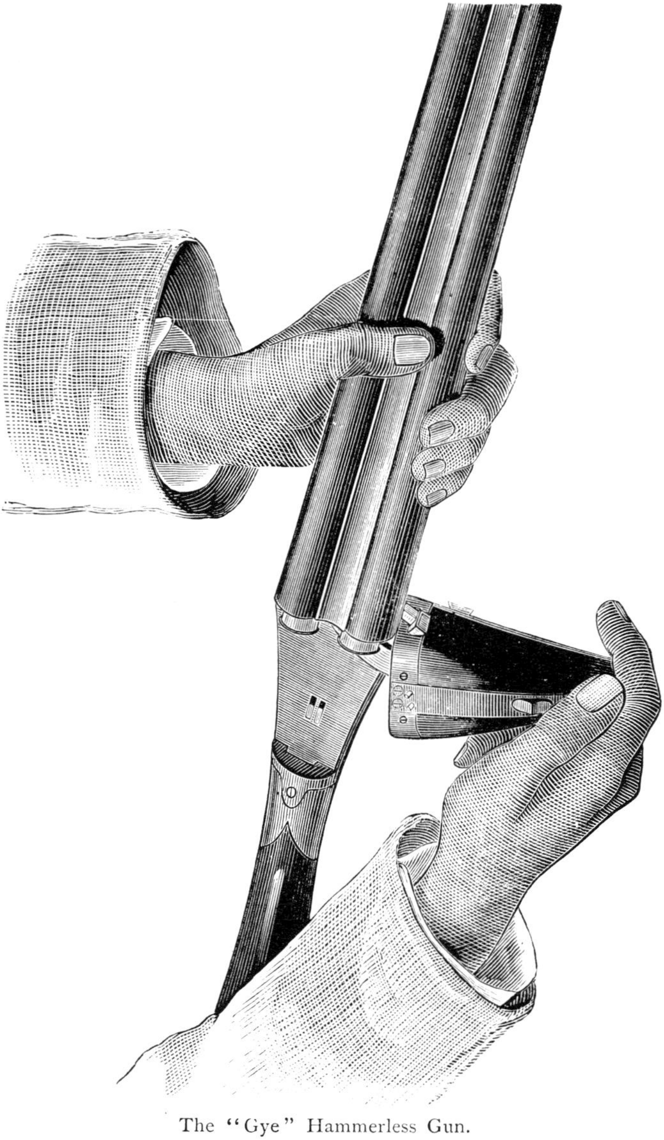

The "Gye" Hammerless Gun.

is cocked by the act of opening the gun; the ejecting locks more nearly resemble the ordinary gun lock, and are cocked as the gun is closed.

The split extractor, one for each barrel, is the main feature of the Needham ejector gun, and is adopted in this principle, but the complication of the parts for the purpose of producing the ejection of the fired case is carried to quite an absurd extent when it is remembered that equal results are obtained with but one lock, and that a strong and simpler one.

GUNS OF DIVERS SYSTEMS.

There is no rule as to the principle on which sporting guns may be constructed. Ingenious gunmakers have devised multifarious plans for effecting the loading at the breech, the cocking of the locks, the ejection of the fired cases; in most cases, however, deviating only in detail from one or other of the typical weapons already described. Among the bolder inventors certain guns have been manufactured and vended which possess widely distinct traits, not necessarily novel.

To enumerate a few only of these: the Lancaster Four-barrelled Gun, in which the barrels are arranged two upon two quadrilaterally, and the cocking mechanism is similar to that of the old Elliott pistol; the Gye Gun, in which the barrels are a fixture and the breech piece is movable, sliding horizontally to the right to give access to the chamber; the Dickson Self-ejecting Gun, in which the lock and extracting mechanism is arranged upon the trigger-plate; the Under-and-Over Gun, in which the barrels are arranged one over the other instead of being placed side by side; double guns with a single trigger; the Dougall breech mechanism, in which the barrels move forward, as well as tilt on the eccentric hinge pin: all of which guns are actually made.

Many details of minor importance are much insisted upon by some manufacturers: such as the nature and position of the safety bolt, on the top, on the side, or beneath. Front or back action locks, &c., all are matters which may with safety be entrusted to the gunmaker, unless the purchaser is used to, or has a predilection for, any special detail of this or like nature. To characterise guns by such feeble traits is beyond the scope of this work; but readers interested will find much additional information on these points in "THE GUN," a work of which mention has already been made.

NOVELTIES.

A gun has been recently introduced in which the lever to operate the breech mechanism is placed in the comb of the stock. This permits of great leverage; it is as quickly manipulated as the usual top lever, and from its length and position permits of greater pressure being put upon the mechanism to withdraw clogged or otherwise lightly fastened bolts. The action spring is stronger, the "bite" on the bolts therefore tighter, and there is but little friction, but owing to the bolt being held in position by a direct spring, it does not keep firmly to its work when the gun is fired.

The appearance of the breech-action is improved by the removal of the lever, but whether or not the gun, as a whole, is improved in appearance sportsmen will decide. The lever is so conveniently placed that the gun is sure to have admirers. The price should be the same as that of other treble-grip guns.

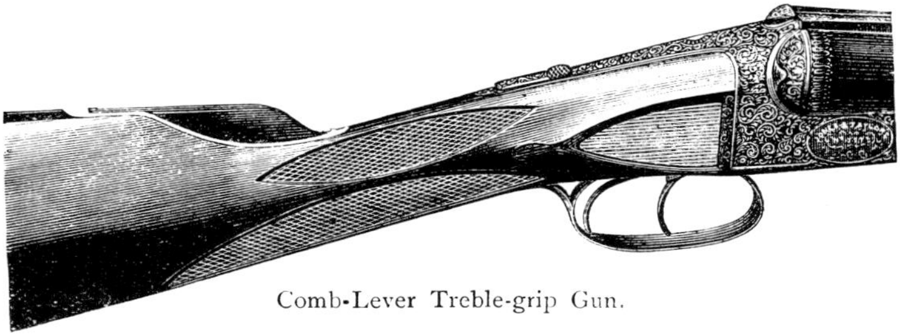

Comb-Lever Treble-grip Gun.

REPEATING SHOT GUNS.

The Repeating Shot Gun is a weapon recently introduced, doubtless with the intention of securing a mercantile success equal to that achieved by the Winchester and kindred magazine rifles. Let it be granted that the repeating rifle is the best of sporting rifles—a point the author will by no means concede—it does not follow that a shot gun constructed upon the same principle will fulfil the requirements of the wing-shot.

Repeating shot guns may be made with an under lever, travelling as does the Winchester, Martin, Kennedy, and other well-known magazine rifles; and such is the Larsen, a gun of Belgian manufacture, but which is not at present popular, nor likely to attain any marked success.

A more popular repeating shot gun is the Spencer, a gun in which the mechanism is worked by the left hand. The fore-end is furnished with a "hand-piece" sliding longitudinally, and actuating a more simple mechanism than that usually found in repeating arms.

The gun can be functioned by the left hand whilst held to the shoulder, and without greatly disturbing the aim.

The well-known shot, Dr. W. F. Carver, attempted to give a "boom" to this gun. He matched himself against time, had six Spencer shot guns, and two assistants to load. Dr. Carver failed, the guns jamming—owing, it is said, to faulty shells. From what the author knows of Dr. Carver, and having supplied him with many thousands of shells and loaded cartridges, he is of opinion that this clever professional shot had cartridges and everything else as perfect as they could be made before he entered upon a trial of such importance.

A public trial of the Spencer shot gun took place in America, and the following sentences are culled from the "Official Report": — "Defective shells were then fired. . . . Result—Slight escape of gas above and below the breech mechanism, but none towards the rear." "Considerable escape above and below, setting paper on fire in one case; no escape of gas towards the rear."

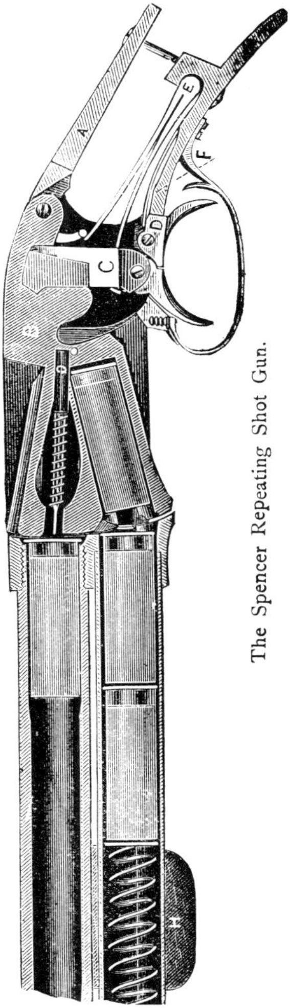

The Spencer Repeating Shot Gun.

The gun was tested for rapidity, irrespective of aim.—"Firer, expert for the Board; time, one minute; eight fired; two thrown out not fired." Magazine loaded before commencing to fire.—"Firer, representative of the gun; one minute; rounds fired, twenty-two. Firer for the Board; time, one minute; fired twelve; thrown out not fired three." Used as a single loader—that is, without calling upon the magazine—eighteen shots per minute could be fired. The ordinary double-barrelled ejecting shot gun can be fired upwards of thirty times per minute, and *it has* been fired and aimed twenty-six times in less than one minute, when the trial had to be discontinued as the barrels had become too hot to hold. The summary states :—

"In all the tests over 378 rounds have been fired from the gun. Of these, *ten* were with defective shells and eight with excessive charges, varying from 120 to 150 grains of powder, with, in several cases, double charges of buckshot.

"The gun remains in excellent condition, as far as its serviceable qualities is concerned, none of the parts being injured or out of order. It has passed very well the various tests to which the Board has subjected it, and the Board is of the opinion that the strength and endurance of the gun are entirely satisfactory.

"In the firing by the expert of the Board, Mr. R. T. Hare, seven cartridges were thrown out unexploded, six in firing rapidly at will, and one in firing for rapidity with accuracy. In the rapid firing by the expert representatives of the gun, three unexploded cartridges were thrown out. This does not include those thrown out because of defective primers, but those cases where the cartridges were not fired because of premature pulling on the trigger, before the primers were in position to be struck by the firing-pin. In the firing by the members of the Board but little difficulty of this kind was experienced."

THE WINCHESTER REPEATING SHOT GUN.

This gun was introduced in the summer of 1887, and has not so far achieved an enviable reputation. The gun is worked by an under-lever, as in the well-known Winchester rifle, but the mechanism is more compact; in fact, the gun is neater in appearance than any repeating shot gun yet introduced, and as a repeater does its work fairly well.

There are perhaps some sportsmen who are inclined to treat the repeating shot guns *au sérieux*. The author cannot do so, for although he admits the mechanism to be ingenious, the results obtained by their use do not warrant their general adoption.

A travelling representative of a firm who manufacture a repeating shot gun, and who was himself expert in the use of the gun, was challenged by a sportsman he met casually to shoot a match with him against time, the sportsman to use a Greener self-acting ejector gun, the expert his repeating shot gun. The result of the match was that the sportsman won by breaking 90 glass balls out of 100 in seven minutes; the expert broke the same number, but occupied more than ten minutes in firing. In this contest the guns became so hot that they had to be cooled by being submerged in a tub of water after every ten or fifteen shots.

SECTION III.—ANALYTICAL.

THE PARTS OF THE GUN.

THE ordinary double-barrelled modern shot gun is composed of 95 pieces, of which the most important are the barrels.

Barrels may be of plain iron, plain steel, or twisted: when they are known technically as scelp-, single-, double-, three-, four-, iron Damascus, stub Damascus, laminated steel.

The plain iron barrel is made of a piece of charcoal iron bent round a mandril, and lap-welded from breech to muzzle between rolls. These barrels are cheap, figureless; and even when perfectly welded and made from the best iron, are quite unfitted for the purpose for which they are used. Many breech-loaders having "decarbonised steel" barrels are

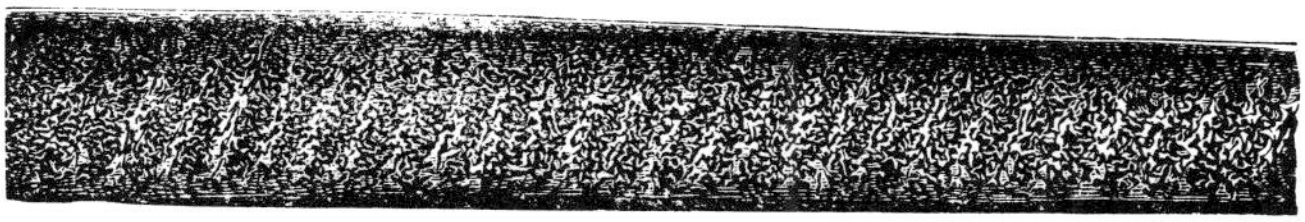

The English Laminated Steel Barrel.

furnished in reality with these worthless tubes. A piece of sheet steel treated in the same way forms a tube still more unsound, but thousands of pairs are made and sold annually; and as there is practically no check upon this fraud on the part of barrel maker, barrel dealer,

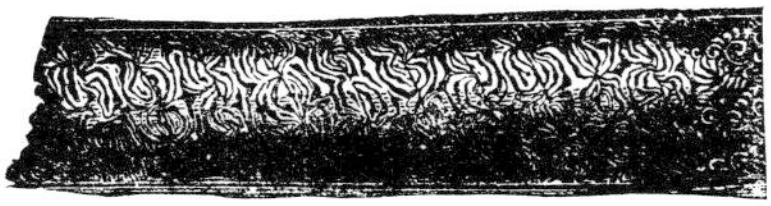

A Silver Steel Damascus Barrel.

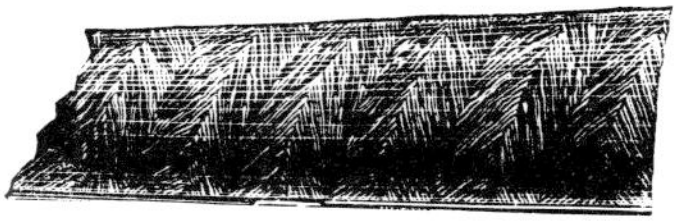

A Plain Laminated Steel Barrel.

barrel filer, gunmaker or gundealer, it is quite possible that such barrels are substituted for the best quality of plain steel barrel. If the sportsman has perfect faith in the honesty of the gunseller, and sufficient faith to believe that none of the persons through whom the barrel has passed would be party to such a fraud, he will buy a gun with plain steel barrel on the seller's recommendation; if he lacks this faith he will purchase a twist barrel, in which the "figure" is a guarantee of its quality. The best solid steel barrels are the "fluid compressed steel" tubes manufactured by Sir Joseph Whitworth's Company. They are very expensive, of uniform good quality, and although they are not, in the author's opinion, equal to

A Fine English Damascus Crolle Barrel.

best twist barrels, he is very pleased to use them at the request of any sportsman requiring them.

Siemens' steel and several other varieties drilled from the solid or

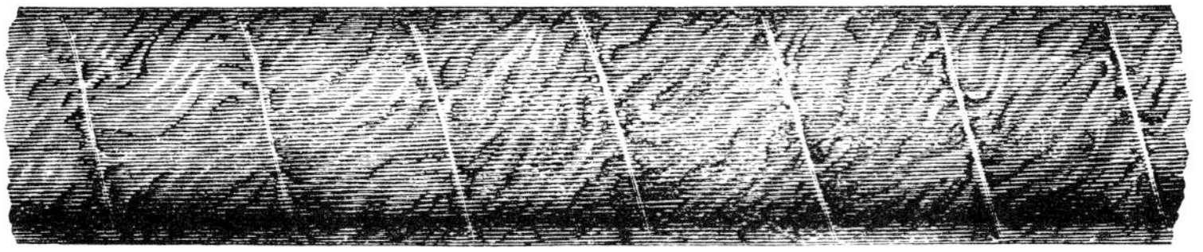

An English Single-Iron Damascus Barrel.

drawn into tubes in the rolling mill, are offered at a less price than the Whitworth barrels, and are often inferior in quality.

Of "twisted" gun barrels, "scelp" is the cheapest. In appearance it

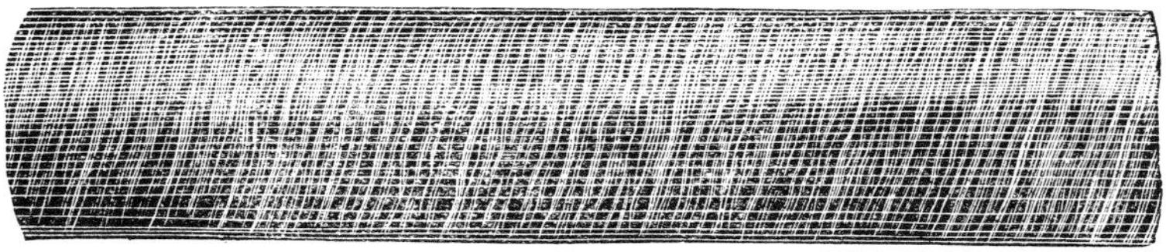

An English Scelp Barrel.

is inferior to either laminated steel or any variety of Damascus. The figure consists of narrow lines, light and dark alternately, which run spirally round and round the barrel from breech to muzzle. It is made from cheap iron, and is only fitted for large heavy barrels.

Damascus and laminated steel barrels have a "cross-figure" in addition to the spiral figure from breech to muzzle, and it is the fineness of this cross figure and the proportion of light to dark lines that will enable even a tyro to distinguish the quality of the barrel.

Fine Stripe Belgian Damascus Barrel.

The illustrations show the chief varieties of twisted gun-barrels.

The barrel, when finished and browned, shows the *grain* of the metal of which it is composed. The darker parts are of iron, the lighter of steel.

Belgian Ordinary Two-iron Damascus or "Boston" Barrel.

Best barrels have as much as eighty per cent. of steel in the metal of which they are composed. The more the metal is worked the better is its

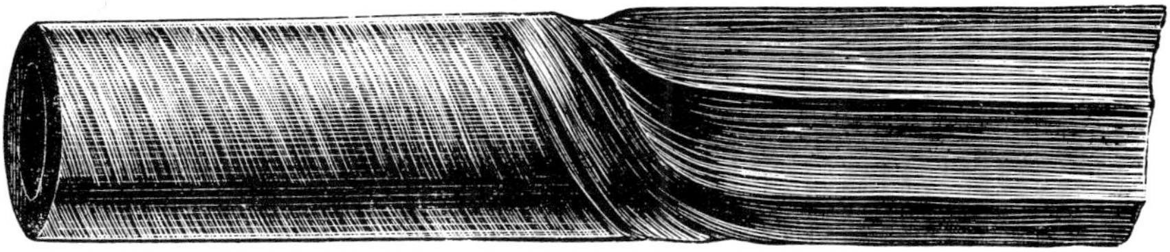

Greener's Solid-Weldless-Twist Gun Barrel.

quality. Only the very best iron will stand the severe twisting to which a very fine Damascus barrel is subjected in the course of manufacture.

Variegated, or "crolle" Damascus is a pleasing but otherwise useless deviation from the true Damascus figure.

Laminated steel gun barrels are welded from a differently composed

The Top and Rib of Greener's Laminated Silver-Steel Barrels.

"gun-iron" than are Damascus barrels, and "the sole difference between steel Damascus and laminated steel" does *not* consist in the manipulation, twisting, and hammering of the rods, as is erroneously stated in the "Modern Sportsman's Gun and Rifle." The difference existing between the various twist gun barrels, and full particulars as to the manufacture of the iron for gun barrels, with much other information bearing on this point, the reader will find in "The Gun and its Development," if he wishes to further examine this purely technical matter.

Another variety of gun barrel, admirably adapted to sporting and military rifles, as well as shot guns, is the author's "solid-weldless-twist" barrel, shown in the preceding illustration.

The grain of the metal runs spirally, as in Damascus barrels, but there are no welds. It has every essential quality of a gun barrel, but can only be produced at a great cost compared with other twist barrels.

The barrels technically known as *tubes* are joined together at the breech by brazing to each other and the barrel *lumps*, at the muzzle by soldering to each other, and at various intervals to *packing*. The top and bottom *ribs* are soldered to the barrel, as is also the *loop* to which the *fore-end* is fastened.

The top rib may be *grooved*, or it may be *flat*; it may be left plain, or it may be *engine-turned*, *file-cut*, *roughened*, or *engraved*, at the choice of the sportsman.

All shot gun barrels are now more or less choke-bored: that is to say, that the barrel is at the muzzle of a less internal diameter than at some point behind the muzzle other than the chamber. A gun barrel constricted to the extent of five-thousandths of an inch is termed a "modified choke;" a full choke is constricted to twenty or thirty thousandths—the larger the bore, proportionately greater must be the constriction. There are two distinct forms of choke: the true choke is obtained by boring the barrel cylinder for nearly the whole length, contracting it at from $2\frac{1}{2}$ to 3 inches from the muzzle. The other plan is to enlarge the bore immediately behind the muzzle, extending the enlargement from 2 to

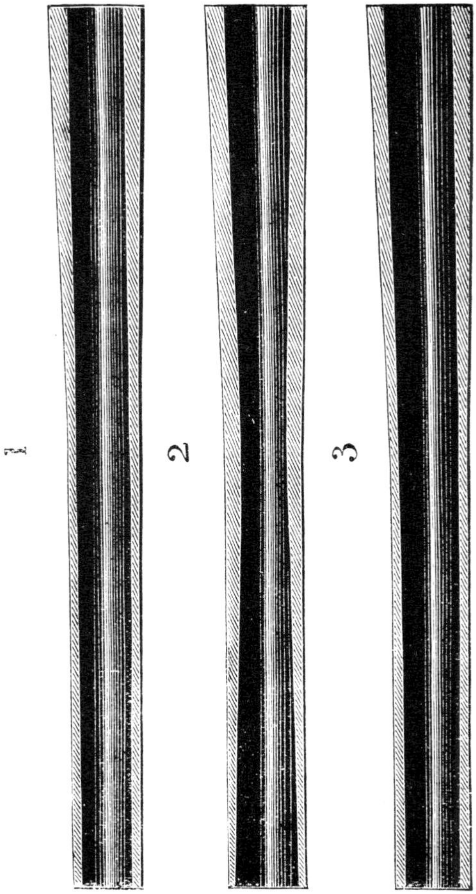

Various Styles of Gun Barrel Boring-"Cylinders."

4 inches towards the breech. A modification of this plan is formed by enlarging the first choke towards the breech in a more elongated form (No. 5); still another modification is made by enlarging the barrels from

the breech to within two or three inches of the muzzle. The No. 6 is the true, or Greener Choke; No. 4, the "Fairburn," or Recess Choke.

A true cylinder barrel (No. 1) is rarely made, the ordinary so-called

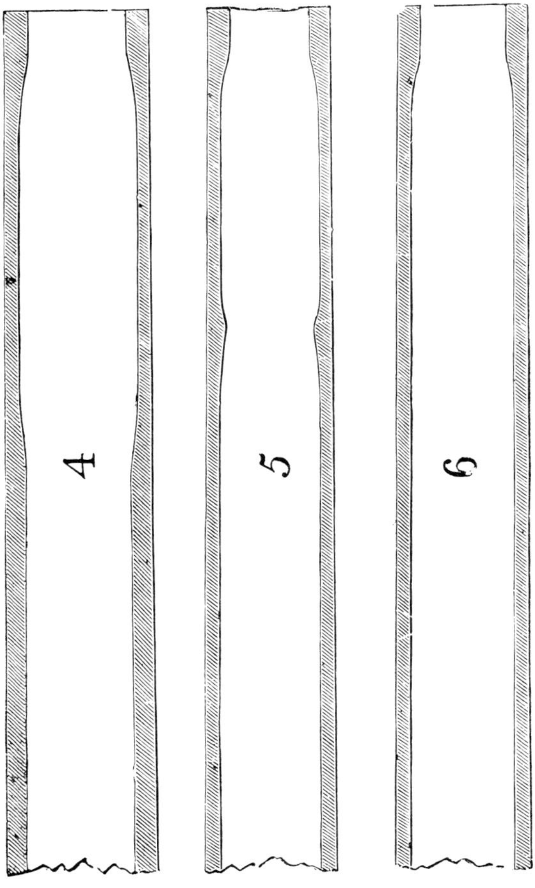

Various Styles of "Choke" Boring.

cylinders being relieved at both ends, as in No. 2, or bored taper, as in No. 3.

There are various degrees of choke. They may be classified as—the full-choke, the half-choke, the quarter-choke, the *improved* cylinder. There is and can be only one true cylinder; the so-called improved cylinder is a barrel slightly choke-bored.

With a 12-bore gun, standard load, distance, and conditions, the ordinary *full-choke* will make an average pattern of	215 pellets.
The half-choke	185 ,,
The quarter-choke	160 ,,
The improved cylinder	140 ,,
The old, or true cylinder	115 ,,

Better shooting than average of 215 pellets can be obtained from an *extra full* choke-bored 12-bore gun.

The chamber is of great importance; it should not be more than ten thousandths of an inch larger in diameter than the smallest case to

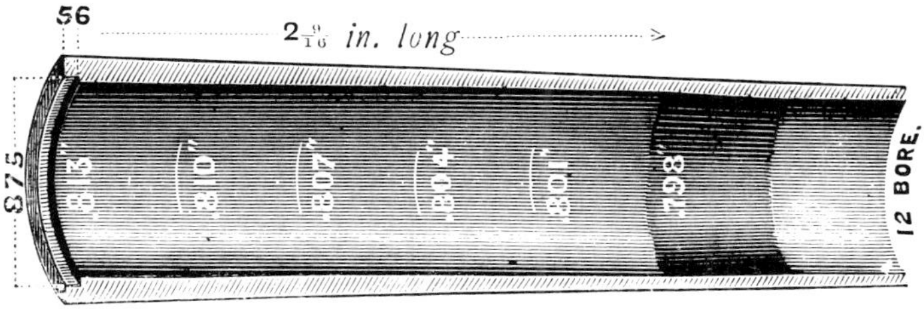

The 12-Bore Cartridge Chamber.

be fired from it; its axis should be one with the axis of the barrel, and it should taper in a gradual cone into the bore of the barrel. The above illustration gives the correct internal diameter of the best 12-bore chamber for brass, paper, English, or other make of cases.

Guns with chambers with sudden cone should be avoided.

The *extractor* is let into the breech end of the barrel, the lower *leg* working in a hole, which should be drilled through the lump, but sometimes faultily drilled in the barrels, to their great detriment. The extractor has sometimes a second *leg*, which in cheap guns not unfrequently works in a hole cut in one or both barrels, and weakens them dangerously. There is no doubt that the majority of gun barrels which burst at the breech have thus faultily fitted extractors.

The *breech-action* comprises the breech-action *body* or *frame*, to which the barrels are fitted, and in which work the holding-down and top *bolts*. The *joint-pin* is that upon which the barrels are hinged, and the *joint* is that turned semicircle against which the *fore-end-iron* abuts. The *locks*

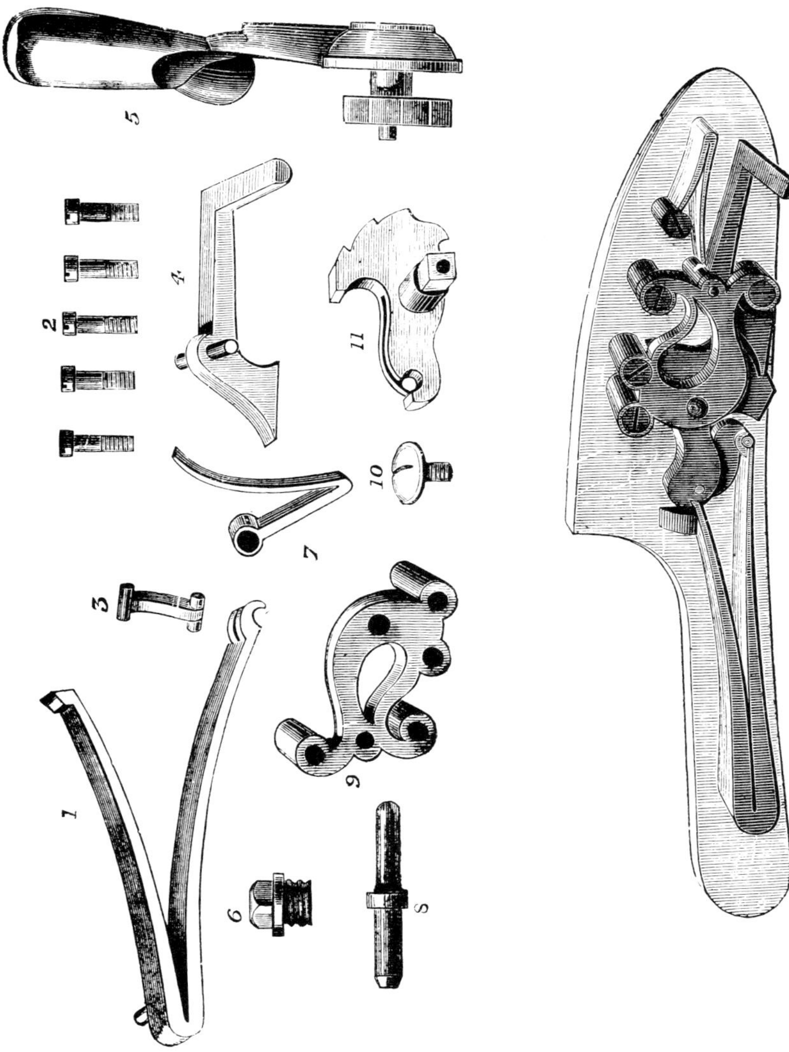

The Modern Rebounding Gun Lock and its Parts.

in a hammer gun may be *front-action* (bar), in which case the mainspring lies in slots cut in the action body, or they may be *back-action*, and wholly let into the stock. A third lock occasionally used is the *solid bar*, in which the appearance is that of a bar lock, but the arrangement of the mechanism that of the back action.

The parts of the ordinary rebounding gun lock are the mainspring (1); the bridle and scear-spring pins (2); the swivel (3); the scear (4); the scear-spring (7); the bridle (9); the tumbler (11); the tumbler-pin (10); the firing mechanism in connection with the lock consists in its simplest form of the hammer (5); the striker (8); the nipple (6).

A different striker is preferred in America: it is as shown below—

The lock mechanism of hammerless guns is sometimes very simple, often complicated. In those guns having side-locks the mechanism is usually similarly arranged to that shown in Section II., with the description of the "Murcott" Gun. When arranged on the trigger-plate, it is too often very complicated. When arranged in the breech-action body, it is found in its simplest form. In the original Anson and Deeley the lock and exploding mechanism consisted of mainspring, tumbler, scear, scear-spring, and pin, and wire pivots for tumbler and scear. This arrangement of the lock mechanism has been generally followed in succeeding patents, and it is only by the use of the Needham scear that this very simple lock can be still further simplified. With the Needham scear no scear-spring is required, the tumbler forcing the scear into bent.

The locks of hammerless guns require some self-acting mechanism which will raise the tumblers to cock, and make the gun ready for firing. The clumsy expedient of a separate lever on the exterior of the gun, to be worked by the hand, is altogether out of the question; and it is not to the credit of world-renowned American ingenuity that a reputable American gunmaker has produced and sold this miserable contrivance, to the exclusion of better mechanisms. The various guns which "cock" by the "lever," the "barrels," or by springs, have already been described, and it will suffice if here is simply detailed the mechanism of one of the most, if not the most, perfect barrel-cocker.

A cocking swivel pivoted in the barrel lump has a hooked extremity, which, engaging with the lock tumblers, carries them to full cock as the barrels are dropped for loading. A sliding stem, having an abutment against the fore-end, keeps the cocking swivel in its position, and permits of the barrels being put into the breech-action body, irrespective of the position of the tumblers: that is, whether cocked or not.

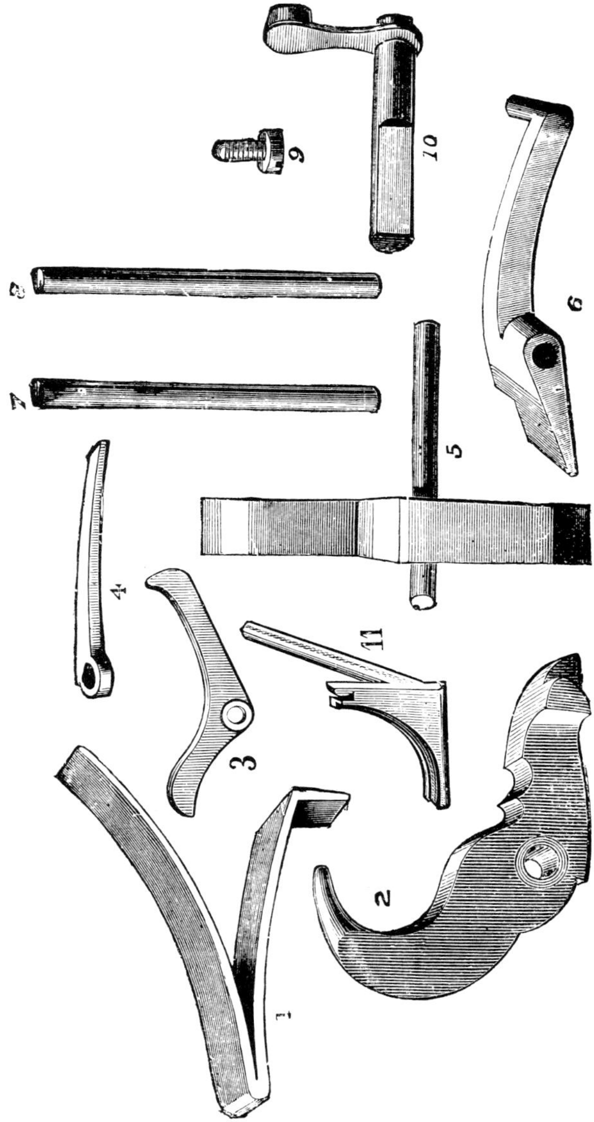

The Lock Mechanism of an Ejector Hammerless Gun.

(1) The mainspring; (2) the tumbler and striker in one; (3) an ejecting lever; (4) scear-spring; (5) a tumbler pivoted; (6) the scear; (7 and 8) tumbler and scear pivots; (9) scear-spring pin; (10) the trigger safety.

SAFETY BOLTS.

Nearly all hammerless guns are fitted with one or more safety bolts, the said bolts being of two distinct kinds—those automatic in their action, and those dependent upon the hand of the shooter to put them to "safe."

Automatic safety bolts are usually applied to the lock tumblers in the shape of a secondary scear or as an intercepting bolt. They are neither more or less than the half-cock of the old muzzle-loading gun.

One of the most popular is the Scott, shown in the accompanying illustrations.

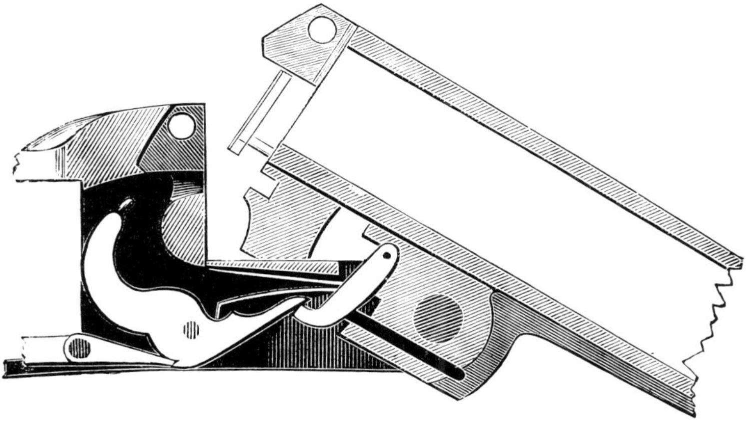

The Cocking Mechanism of W. W. Greener's *Facile Princeps* Hammerless Gun.

A lever is pivoted so that its one extremity, C, comes into contact with the trigger exactly as does the scear, D. A projection, A, on the other extremity of this lever will, under certain conditions, block the tumbler, B, so as to prevent its reaching the exploding pin, E. In the three figures the lock is shown cocked and ready for firing in the first; in the second the trigger has been pulled, and the tumbler released and struck the striker, E. In the last it is supposed that the tumbler has been liberated by some means other than the pulling of the trigger, and the tumbler has consequently failed to reach E, being effectually blocked by the stud A. This safety, as made, is not strong enough to be relied on implicitly.

A secondary scear, working exactly as, but quite independently of, the ordinary scear, effects the same purpose, and is used by the author on some guns, also on guns made by others.

As this order of safety provides in no way for accidental discharge by

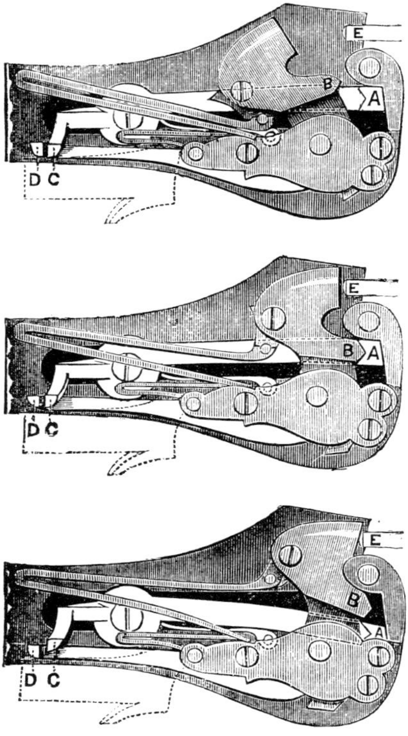

Scott's Automatic Intercepting Safety Bolt.

accident to the trigger or from inadvertently touching it, a trigger safety is used in addition to the intercepting safety, and this may be either automatic or independent in its action. The common form of trigger safety is a T bolt sliding on the top of the hand of the gun thrown into "safe," by gearing with the holding-down bolt on the opening of the gun, and requiring to be pushed forward by the shooter before the gun can be fired. It requires a slot through the hand of the stock at its weakest point, and is certainly better dispensed with.

The Greener side safety bolts the triggers, and is independent in its action, although it may be made automatic if so desired. It is shown on page 24, and neither weakens the stock nor detracts from its appearance. In action it is most effective. The safety consists of a round rod which has a flat filed upon it, the said flat side being so placed that the trigger slips by, but as soon as the rod is turned five degrees they are held down by the rod itself.

The old grip safety of muzzle-loading days, of so little use because the gun when carried is usually gripped on the safe, and the triggers consequently left free, has been resuscitated, and in several forms is used by various gunmakers. The "Silver" safety is on this principle, but with gearing so modified that not only are the triggers bolted, but the scears and tumblers also, as they are all unbolted immediately the gun is grasped.

With the muzzle-loaders a shoulder safety bolt was also used sometimes, consisting of a movable heel-plate connected by a rod with a safety mechanism. As soon as the gun was pressed to the shoulder the triggers, &c., were released, but as this was also the case when the gun was placed on the ground for loading and ramming home the charge, it was of little use as a safety. In 1879 the author introduced a certain modification for

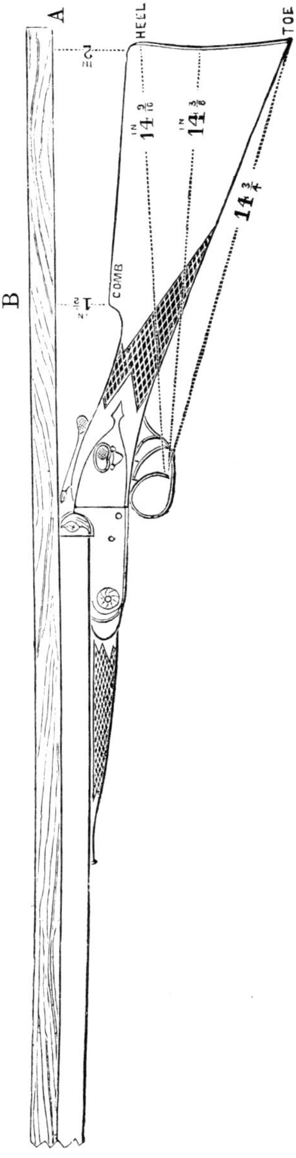

The "English" Gun Stock.

The "American" Gun Stock.

hammerless breech-loading guns, but it did not have much popularity. It has since been several times re-invented by ambitious makers.

There are various complicated mechanisms misnamed safeties, which would require more space to describe than they merit, and they should certainly be studiously avoided by all sportsmen who regard their own safety and that of their companions.

The dimensions and shape of the *gun stock* have been discussed amongst sportsmen and gunmakers for many years, and at present there is no definite authority for having the gun stock shaped to any of the prevalent fashions.

The measures of the gun stock, including the *bend* of the gun, are of the utmost importance to the user of the gun, and must *suit* his particular

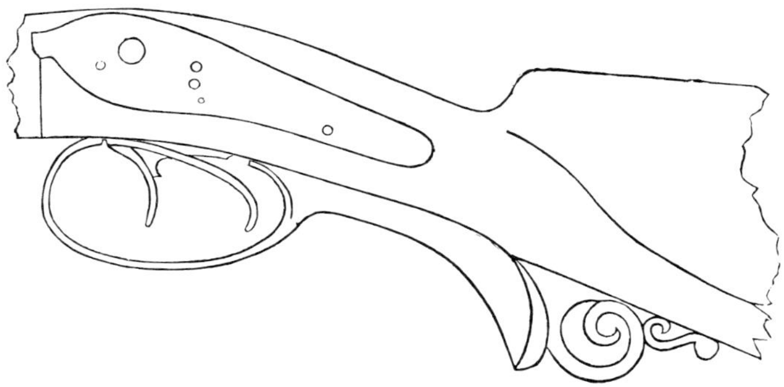

The German Horn Grip Guard.

method of handling the gun, as well as the stock being of such dimensions as the shooter's *build*—*i.e.*, length of arm, breadth of chest, &c.—may determine.

The adjoined figure shows an outline of the ordinary sporting gun stock, as used by the majority of British sportsmen. The measures of a gun stock may be ascertained as follows :—

Take a piece of wood or iron with a perfectly straight edge, sufficiently long to reach from the sight on the muzzle to the extremity of the butt; lay this straight-edge along the rib, and measure the distance from A to HEEL, and from B to COMB. This is the BEND. The LENGTHS required will be from the centre of the fore or right-hand trigger to the HEEL, CENTRE, and TOE respectively, and the depth from the heel to the toe. The circumference of the hand may be obtained by passing a string round it *immediately* behind the trigger-guard, measuring the string. In taking the length, measure the extreme length, and not to the edge of the heel-plate. The dimensions given above are in due proportion, and as usually made for English and American sportsmen respectively.

CAST OFF is the amount the stock is thrown out of truth with the barrels in a *lateral* direction (*vide* p. 51). Most gun stocks are twisted over—that is, the toe of the butt is more "cast off" than the heel—the usual "cast off" is 3-16ths for heel, and 3-8ths for toe. The object of "cast off" is to admit of a proper aim being easily taken, and the amount will vary according to the build and physique of the person for whom the gun is constructed.

Balance.—This is always to be measured from the breech ends of the barrels. It is best to balance the gun on thin string.

There are other shapes of gun stocks, modifications of these two

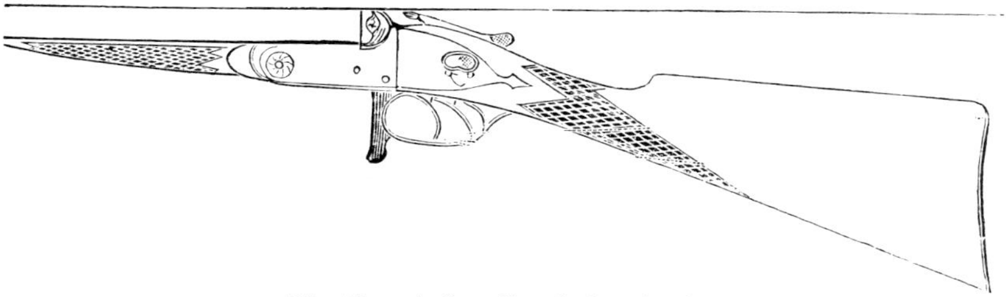

The Horn before Guard Gun Stock.

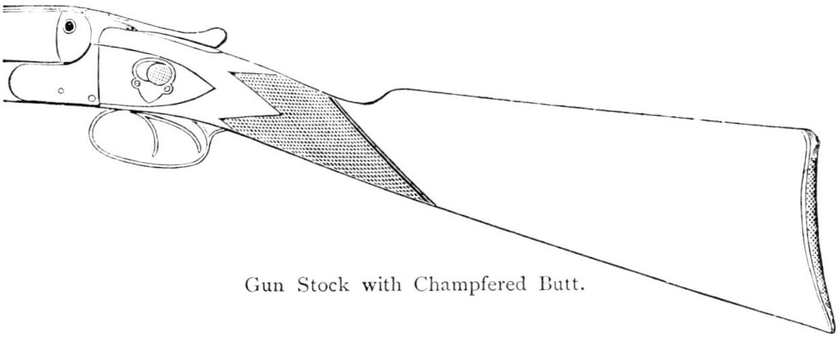

Gun Stock with Champfered Butt.

types, with which many sportsmen are acquainted, but to others they will be novel, and may offer certain advantages. First, there is the *horn guard*, equivalent to the *scroll guard* of the old-fashioned English rifle. This guard is supposed to allow a better and firmer grip of the gun to be obtained with the right hand—the same advantage as claimed for the *pistol hand* stock, and it moreover prevents the fourth finger of the right hand from being bruised by the back of the trigger-guard.

The horn guard is much used by some Continental sportsmen, and the German gunmakers particularly, fashion it into an ornamental fitting for either the shot gun or rifle.

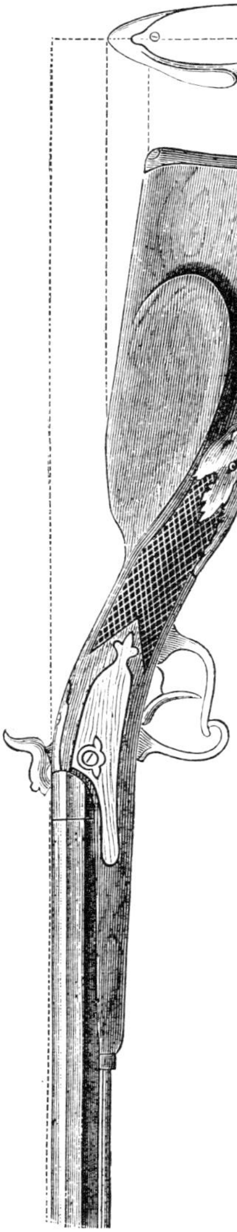

The Stock of a Belgian Muzzle-loader.

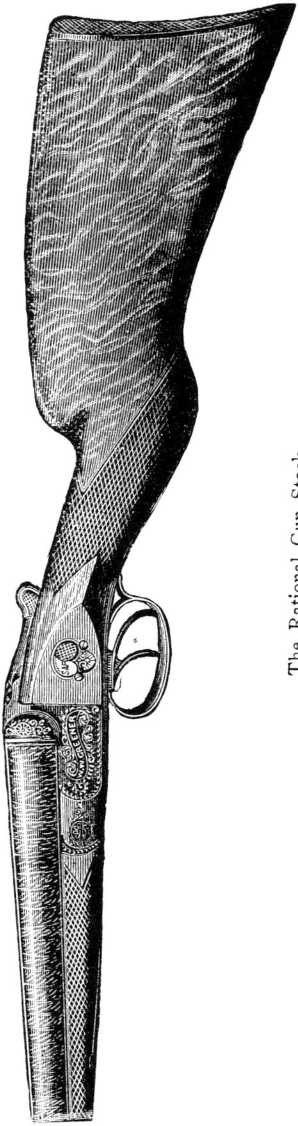

The Rational Gun Stock.

Another Continental form is the shield guard, or horn before guard. With this style of stock the gun is grasped just in front of the trigger guard by the thumb and forefinger of the left hand, the palm of the left hand and the remaining fingers being firmly pressed against the guard. This style of holding the gun is not to be commended, but it must be admitted that many fine shots are to be found who never hold their gun differently. The "Shield" guard is shown on the illustration of the Belgian muzzle-loader.

Sometimes guns are made with a stud, handle, or projection screwed perpendicularly into the fore-end, and which is grasped by the left hand. With this stock the gun lies upon the clenched fist. It is not to be recommended, nor is it much used; its use, however, sometimes corrects the very bad fault of grasping the gun with the left hand at various distances from the breech, a practice injurious to good or even fair shooting.

The lengths of the gun stock from fore trigger to toe and heel will regulate the angle of the butt, and the *cast off* will throw the butt over a little, so that unless the butt were rounded or champfered, its *edge* only would touch against the shoulder. The amount of champfer required will depend upon the amount of cast off, and the build of the person for whom the gun is intended. Dr. W. F. Carver always shoots with a heelplate not only much hollowed—*i.e.*, *very much* shorter to centre than to the extremities—but also champfered so as to fit squarely against the muscles of his shoulder. Many shooters will find it more comfortable to shoot with a gun having the butt so rounded or sloped than with the usual butt, which is of equal length to either edge.

Another form of gun stock is that much used on cheap guns for South American sportsmen. The style is not unlike that of the Continental guns of the latter end of last century, and there are many first-class muzzle-loading guns of French make which have the stocks fashioned to this model, which handle well. It must be admitted that they have the appearance of a clumsily finished weapon, and few English sportsmen would think that quick and accurate shooting could be got from a gun with such a stock, but a trial of a well-balanced weapon having this style of stock will convince any one that this shape of the stock is conducive to rapid aiming. The rational stock has been recently introduced by the author, and it will, he thinks, be found to embody qualities long sought in pistol hand and very straight stocks.

In this stock, as will be seen from the illustration, there is more than the usual bend at the *bump* or *heel*, and that the comb is not straight, but *arched* slightly; and as the cheek touches the stock about midway between the *heel* and the *thumb*, it is there, and there only, that the stock need be straight.

With the usual English gun stock, put up in the usual manner, it will be found that about *one-third* of the upper part of the *butt* projects.

Shooting a Straight-Stocked Gun.

EXPLANATION.—At the time of firing the whole of the butt-plate should be in contact with the shoulder of the shooter. When firing at ground game this is almost impossible with a straight gun stock having the usual amount of toe. Usually the gun stock projects much above the shooter's shoulder—as shown—the result being that the recoil causes the gun to jump up. If the gun be fired whilst held as shown above, the cheek of the shooter would have to sustain a goodly share of the force of the recoil.

Shooting with a Rational Stock.

EXPLANATION.—The rational stock is, at the time of firing, squarely bedded against the shoulder of the shooter, and whether firing at ground game or at birds flying straight over, must always have a greater portion of its butt-plate in contact with the shoulder of the shooter than it is possible to get when using the ordinary straight stock.

above the shooter, and has *no* bearing against the shoulder. This leaves the sharp, narrow toe to steady the gun and to take the recoil.

With the rational stock the face of the shooter will be resting upon the stock when the *bump* or *heel* has reached a level of the shoulder, and the whole of the *butt* will find a bearing in the hollow of the shooter's shoulder.

The bend of the gun will, with the "rational" stock, be about $2\frac{1}{2}$ inch at *heel,* $1\frac{1}{2}$ at *comb*, and $1\frac{5}{8}$ midway between heel and comb.

In the rational stock the wood is left *thicker* at the toe, and thinner than usual at the heel. This enables the shooter more easily to get his gun into proper position—with the large heel and thick comb the stock often catches against the shoulder, and *jumps* afterwards to the position from which it is fired. It is believed that by reversing the taper of the butt the gun will *invariably* be brought with greater certainty and speed to its proper bearing at the shoulder.

The rational gun stock also, instead of being straight from trigger bow to toe, is arched slightly near the end of the grip. This *roaching* of the stock is a very modified form of pistol grip, and allows not only of a firmer and easier grip being taken, but also permits the elbow to fall lower and to a more natural position when the gun is at the shoulder and about to be fired. The gun stock must be so fashioned that the butt shall be at a right angle, or nearly so, to the barrels, and the gun will stand with the barrels *almost* perpendicular. Some, however, prefer that the gun when stood upright shall be such that the sight and the centre of the butt shall be in a plumb-line.

The cast off, as already explained, is the *lateral* deviation of the stock from the common axis of the barrels. This permits of the barrels being more readily aligned.

In the gun with the *cheek-piece* the cast off of the gun is almost, and frequently quite annulled by the projection on the left side of the stock called the cheek-piece.

From the dotted lines in the illustration indicating the full centre it will be seen that the stock has an advantage to the right, but this advantage is compensated for by the projecting cheek-piece, which at the centre of the stock actually projects beyond the true line.

The use for and necessity of *cast off* will at once become apparent on an examination of the next illustration, showing a gun so cast over that it may be aligned from the right shoulder with the left eye. This kind of stock serves a very useful purpose. Unfortunately, too many shooters lose the sight of the right eye from some mishap when using their guns, and to such a gun with a stock of this description is an absolute necessity. But more than one style of stock has been devised for these sportsmen, and the second model shown is of the two to be preferred ; it is quite as handy and strong, and gives the same shaped comb at the same angle as an ordinary stock. The "Monopeian" gun comes into this same category, although the result obtained is not by bending over or so fashioning the stock that

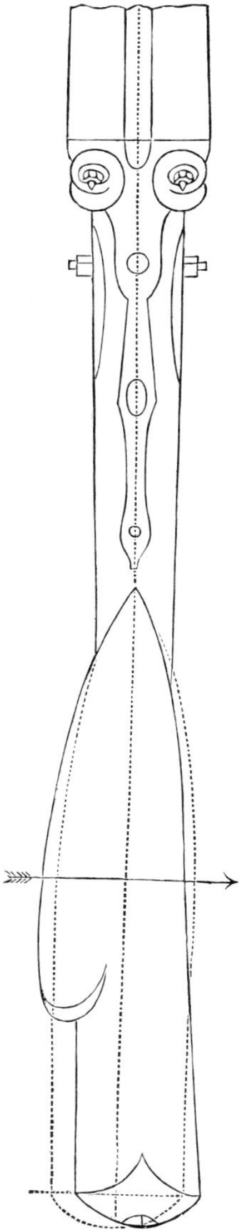

Gun Stock with Cheek Piece.

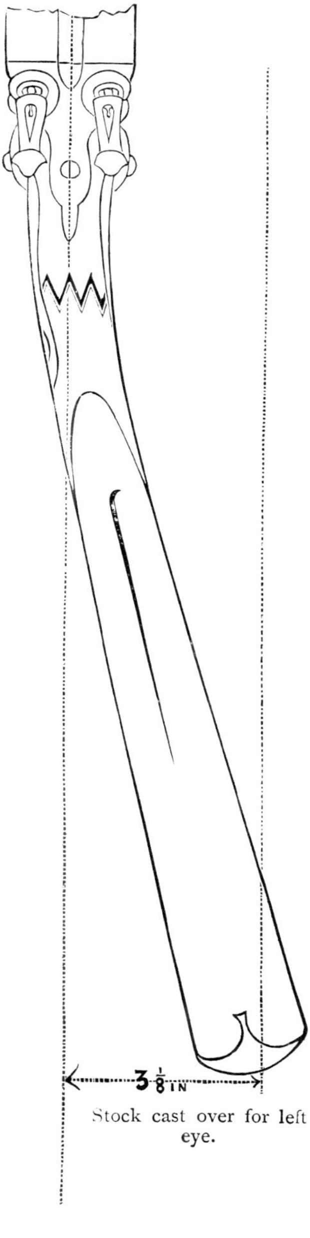

Stock cast over for left eye.

the left eye may see over to the rib and align the gun, but the sight is brought out to the left side of the left barrel, and an additional sight affixed to the breech.

The illustration will give an accurate idea of the principle of the "Monopeian" gun, which is the invention of the Rev. E. Elmhirst, B.A., of Showell, and is used by many monopeians, who can testify to its utility for the purpose for which it was intended. The disadvantages arising from firing a gun with the stock bent over for the left eye are obviated by this gun. With the very much cast off stock the left barrel must fire to the left, and the gun will handle clumsily to any one who has been used to the ordinary gun stock, but a single gun will not appear so clumsy in this respect as a double-barrelled one.

The "Monopeian" gun has two sights attached to the left barrel, one at the breech, the other at or near to the muzzle. These sights project about one inch from the barrel, as in the illustration, and the back sight need not be more than 10-sixteenths, and the fore sight more than one half-inch above the top of the barrels if the stock is properly constructed.

The *stock* of the gun must be of well-seasoned walnut, as beautifully marked as can be found. The *furniture*, that is, *triggers*, *guard*, heel-plate, escutcheon, &c., fashioned to taste.

The *fore-end* will also be of walnut-wood, and should be readily detachable from the barrel. Either of the illustrated bolts will permit of this, and they will at once be recognised.

ADJUNCTS TO THE BREECH-LOADING SHOT GUN.

There are various devices occasionally added to the shot gun with some specific object, as the fitting of the india-rubber butt-plate, to lessen the effect of the recoil. This well-known appliance has never been equalled by any combination of steel springs, and the most recent development, a combination of solid rubber and hollow rubber pipes in the composition of the butt-plate, does not seem to offer any advantage over the original "Silver's anti-recoil pad," which cannot, however, be recommended in preference to a properly fitted stock, with which, if the gun be firmly held, the recoil will not be felt to be heavier than with the pad.

A pad to draw on over the butt-plate is provided with an india-rubber sponge cushion, which, coming between the butt of the gun and the shoulder of the shooter, in some measure lessens the effect of the recoil, but the arrangement can only be considered as a temporary one.

A cheek-pad has also been designed, which, drawn over the stock, as in the annexed drawing, will lighten the blow on the cheek given sometimes by a gun which fits badly or is over-loaded. The pad is self-adjusting, and by altering its position on the stock the fit of a gun may be altered.

The fault with the pad, as made, is that it completely neutralises the cast off of the gun.

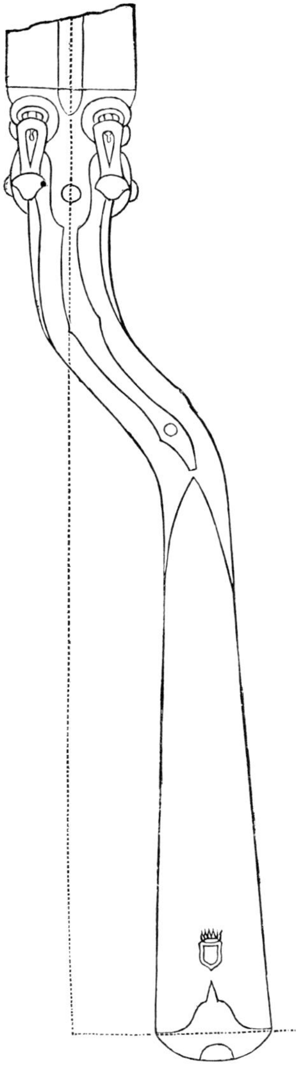

Bent Stock for use from right shoulder with the left eye.

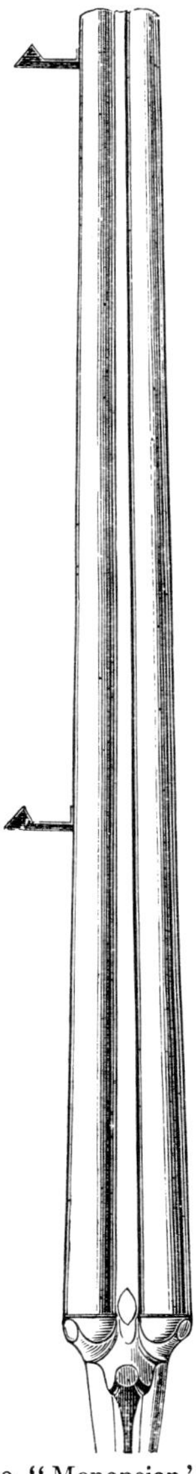

The "Monopeian" Gun.

To prevent the bruising of the forefinger from the recoil of the upright trigger, the right-hand trigger is sometimes hinged, and is pushed forward by the forefinger as the gun recoils. A more simple way is for the gun to be built with wider spaces between the triggers, but this may increase the difference in the length of stock when pulling the right or left-hand trigger. The trigger may be covered with soft leather, or with an india-rubber inflated cover, or an inflated india-rubber or leather guard may be worn

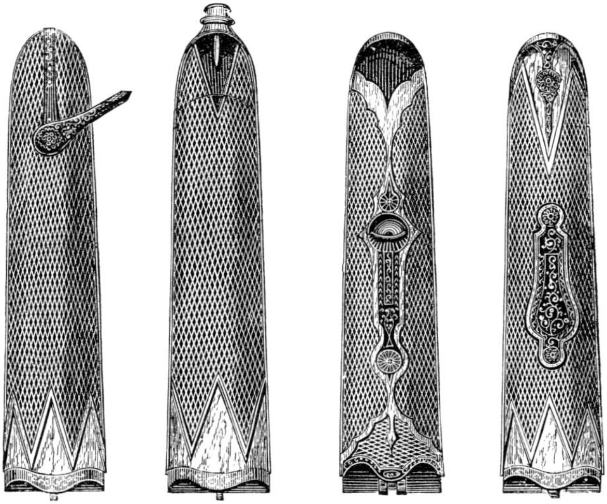

Fore-ends.

upon the forefinger. Occasionally the third finger is bruised by the back of the bow of the trigger-guard, and it may also be protected by a guard, or a leather, horn, or other stop may be fitted to the trigger-bow.

The *sight* of the Shot Gun is of little importance; no gun certainly looks complete without one. To some the Gilbert Shooting Corrector will prove advantageous.

The *finish* of a gun, although generally meant to understand the amount of ornamentation in the way of chequering and gloss upon the wood-work, in reality means something of much greater importance. The well-finished gun will not only have a beautiful exterior, but the barrels, every piece of the mechanism, every bolt, pin, and screw, will not only be perfectly fitted and well polished, but so placed as to be of actual service. A well-finished gun will be well balanced, well bored, will work smoothly, and shoot well, and should have had the greatest care bestowed upon it in every stage of manufacture, and the greatest attention given to every detail. Finish in this sense cannot be over-rated; finish, as meant by the mere scratching and polishing of the exterior, is of less moment.

Sometimes it is convenient to have a sling to the gun, in order that it

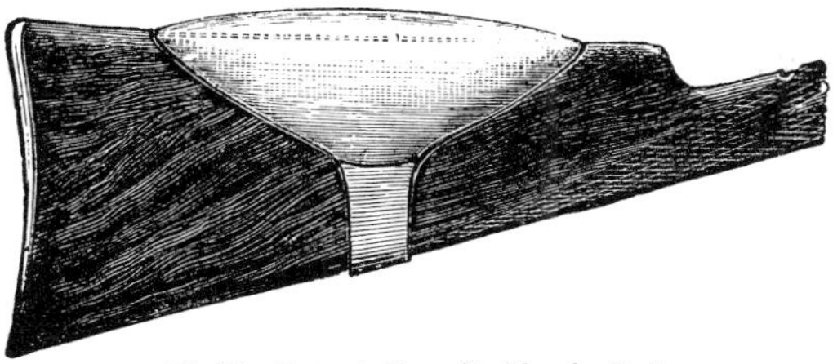

Hall's Patent Recoil Cheek Pad.

may be slung across the shoulder. For this purpose there are attachable slings made, but it is neater to have soldered upon the barrel and screwed into the stock a swivel for a flat strap, or an eye; if the latter, a swivel will have to be attached to each end of the strap used as a sling.

In gunmaking there are a thousand and one little things to be done,

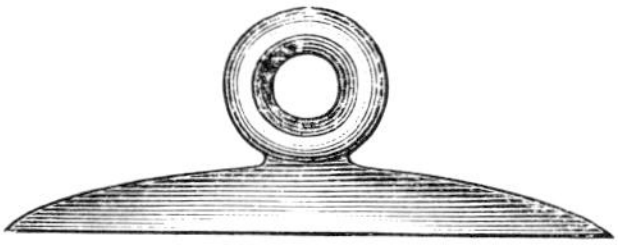

Gun-barrel Eye for a Swivelled Sling.

A "Swivel" for Flat Sling.

which, of little importance in themselves, make all the difference between a good gun and a gun that will shoot.

In "THE GUN AND ITS DEVELOPMENT" the various stages of gun-making are dealt with at length, and a perusal will convince any reader that a gun is not an article that can be produced for a guinea—the present price of a commodity sold for and commonly known as a "gun."

SECTION IV.—TECHNICAL.

CALIBRE.

THE breech-loading sporting Shot Gun is made in the following gauges: 28-diameter of the interior of the barrel being ·550 inch, 24-bore ·579, 20-bore 615, ·16-bore ·662 inch, 14-bore ·693, 12-bore ·729, 10-bore ·775, 8-bore ·835, 4-bore 1·052, 2-bore 1·325 inch. The nomenclature by "bore" used by gunmakers is based upon the size of the bullet mould. A mould casting fourteen spherical bullets to the pound would be a fourteen-mould; the barrel that bullet would fit would be a fourteen-bore. The gauge is determined by the size of the cartridge cases. The calibre accurately means the diameter, but the three terms are commonly used indiscriminately.

WEIGHT.

A 28-gauge double gun with 26-inch barrels should weigh under 4½lbs.; a 24-gauge with 28-inch barrels should weigh 4¾lbs.; a 20-gauge with 30-inch barrels should weigh 5½lbs.; with 28-inch barrels 5¼lbs.; with 26-inch barrels 5lbs.; with 24-inch barrels, and built as a miniature weapon, 4lbs. 12ozs.; a 16-gauge with 30-inch barrels may weigh 6¾lbs., and be made a powerful weapon, but 6½lbs. is the full average weight; if with 28-inch barrels 6¼ or even 6lbs.; with 25-inch barrels, and built as a miniature gun, 5lbs. 2ozs. The fourteen-gauge is seldom used; if with 30-inch barrels a double gun should weigh 6¾lbs., with 28-inch barrels 6½lbs., and a miniature 14-bore gun with 24-inch barrels should weigh 5lbs. 10 ozs. The 12-gauge ordinary game gun with 30-inch barrels should weigh 7lbs., not over; if with extra full choke-bored barrels, and intended for use with full loads, 7¼lbs.; the 12-gauge 30-inch or 31-inch, for pigeon shooting, 7½lbs.; the 12-gauge 32 for heavy loads, shore shooting and general utility gun from 7¾ to 8lbs.; the light 12-gauge game gun, with 28 or 30 inch barrels, 6¾lbs.; the short-barrelled covert gun, 6½lbs.; the miniature 12-gauge gun, with 27-inch barrels, and every part proportionately reduced, 5¾lbs.

The 10-gauge 30-inch general Trap and Duck gun 8½lbs.; with 32-inch barrels for duck shooting especially, 9lbs.; with 32 or 33-inch barrels, for heavy charges in long paper or brass cases, 10 to 10½lbs.; light short-barrelled 10 bores for making a large killing circle at moderate ranges, and for use with large shot, 7¾lbs.

The 8-gauge from 12 to 15lbs., the 4-gauge from 15 to 18lbs., the 2-gauge 18 to 20lbs. There is also the miniature 8-gauge, with 32-inch barrels, 11lbs.

Single guns of 16-gauge may weigh about 4lbs., 12-gauge $4\frac{3}{4}$lbs., 10-gauge 7lbs., 8-gauge 11lbs., 4-gauge 15lbs., 2-gauge 18lbs.

The weights given above are the average and usual weights, and are proportionate to the calibre and length of the barrel. Well-balanced guns can be made of other weights, but would require to be specially built, with furniture and fittings of special size.

SHOOTING.

Shooting is technically the pattern and penetration of the gun at a certain distance with a certain charge. Gunmakers invariably shoot at a 30-inch circle at 40 yards distant from the butt of the gun, and would technically state a gun's shooting thus--L. 200, R. 195. That is to say, that the gun with the *standard charge* for its gauge has made, or is required to make, an average pattern of 200 pellets with the left, and 195 with the right, inside of the 30-inch circle, at 40 yards distance. The size of shot is 270 to the ounce—the English No. 6 size.

The following are the standard charges of the various gauges :—

GAUGE.	POWDER. Drams.	SHOT. Ounce.	CARTRIDGE. Long.
·360	$\frac{1}{2}$	$\frac{3}{16}$	$1\frac{3}{4}$
·410	$\frac{3}{4}$	$\frac{3}{8}$	2
32	$1\frac{3}{8}$	$\frac{9}{16}$	$2\frac{1}{2}$
28	$1\frac{1}{2}$	$\frac{3}{4}$	$2\frac{1}{2}$
24	$1\frac{3}{4}$	$\frac{7}{8}$	$2\frac{1}{2}$
20	$2\frac{1}{4}$	1	$2\frac{1}{2}$
16	$2\frac{1}{2}$	1	$2\frac{1}{2}$
12	3	$1\frac{1}{8}$	$2\frac{5}{8}$
12 pigeon gun.	$3\frac{1}{4}$	$1\frac{1}{4}$	$2\frac{3}{4}$
10	4	$1\frac{3}{8}$	$2\frac{7}{8}$
8	7* 6ϕ	$2\frac{1}{2}$	$3\frac{1}{4}$
4	*9	$3\frac{1}{4}$	4
2 ϕ	10	$3\frac{1}{2}$	$4\frac{1}{2}$

* Brass Cases. ϕ Paper Cases.

All paper cases vary considerably in length, the 12-gauge being often $2\frac{9}{16}$-inch.

There is no generally accepted method either of determining or stating the penetration of guns. Penetration may be given as the average striking force at impact of each pellet as ascertained by the "field" force gauge, or in the penetration of sheets of brown or other paper, of straw-boards, or the depth of soft pine pierced.

THE PROOF OF GUN BARRELS.

Since 1637 the proof of English-made guns has been more or less directly under the control of the British Legislature. By Royal Charters,

Acts, and Amendments, it has been sought to prevent the issue of faulty weapons; and although, necessarily, no rule or test will be equal to every case, it cannot be doubted that the obligatory proof has acted very beneficially. Acts of Parliament, scales of proof charges, and all regulations and rules possibly applicable and efficient at the time of their compilation, are not likely to long remain so in these days of invention.

Unfortunately, the Proof House authorities have hitherto been slow to realise that changes of great importance to the users of guns are not provided for by their rules and regulations, and were it not for constant agitation, it is very possible that the gun of 1888 would be proved with the charge and kind of explosive ordered for that of 1868. Fortunately, there are some gunmakers who, noticing new inventions or the introduction of new explosives, clamour for such alterations of the Proof House rules and charges as will meet the case.

The new rules of proof which have just come into operation provide for a more efficient test of shot guns and sporting rifles, and there is every reason to believe that guns proved under the new scale will be safe with the standard charge of any sporting gunpowder.

As to breech-loading shot guns, the barrels shall be twice proved, the first proof being only a provisional test of the barrel tubes, the second, or definitive proof, after the breech action and barrels have been properly shaped and fitted and finished filing.

Muzzle-loading shot guns may be twice proved, or proved in the finished state with the heavier charge, according to the scale for provisional proof. Single-barrelled muzzle-loading guns with plain iron barrels are proved once only. The gunpowder used in proof shall be a mixed grain rifle powder, of quality equal to that used by the British Government, but a *supplementary* proof of any gun may be made with any explosive, if so desired.

The bullets of soft lead of the specific gravity of 11·352, and for rifled choke-bore barrels are conical in form, and of a diameter not more than ·005 of an inch less than the muzzle diameter of the barrel.

The shot used shall be of soft lead, and the size No. 6. The wads of solid felt, and not exceeding in thickness one diameter of the bore—one wad only over powder and one over bullet.

The marks of proofs are stamped upon the barrels in accordance with the following rule:—Shot guns and rifled choke-bore shot guns have the provisional proof mark struck on the round of the barrel near the breech end, and the definitive proof mark and view mark shall be impressed upon the barrel above the provisional proof mark, and if the barrel be constructed with a patent breech, or with a breech-loading action, or with a breech block or chamber, the view mark shall be also impressed upon the patent breech, breech-loading action, shoe, breech block, or chamber, with which the barrel is connected.

On double breech-loaders, the barrels for which have been submitted for provisional proof before being put together, the provisional proof mark shall be impressed upon the round of the barrels near the breech, in front of the flats, or in a corresponding position where there are no flats, and the definitive proof mark and view mark shall be impressed upon the flats or in a corresponding position when there are no flats. If provisionally proved after being put together, the provisional proof mark shall be impressed upon the flats at the breech end, or in a corresponding position, and the definitive proof mark and view mark shall be impressed immediately following the provisional proof mark. The view mark shall also be impressed upon the action or shoe, breech block or blocks, chamber or chambers with which the barrels are connected.

On the barrels the gauge size of the barrel shall be impressed at the definitive proof, and on all barrels for breech-loading shot guns the chamber gauge size, together with the letter "C," signifying Chamber, shall also be impressed following the definitive proof and view marks, and shall be enclosed in a device, thus :—

In all smooth-bore and choke-bored barrels, from 4 to 10 gauge inclusive, the gauge shall be divided into three parts, and be marked accordingly : thus the divisions of gauge 8 would be marked 8, 8-1, 8-2 ; from 11 to 17 inclusive, the gauges shall be divided into two parts, thus divisions of gauge 12 would be 12, 12-1 ; all lesser gauges shall be marked as set forth in the schedule hereto. In all smooth-bore and choke-bored breech-loading barrels the gauge size of the barrel shall be taken at a point nine inches from the breech end. In all other barrels the gauge size shall be taken at the muzzle.

On choke-bored barrels the additional mark "Choke" shall be impressed, and on rifled choke-bored barrels the letter "R" shall be impressed immediately preceding the work Choke, thus—"R. CHOKE."

All barrels of 10 gauge or less, having chambers of three inches or longer, shall be proved with one-sixth more powder than the ordinary definitive proof charge, and the barrel so proved shall be marked thus, and barrels proved in a supplementary proof, with "Schultze," or other special explosive, shall be marked with "Sch." for Schultze, or with a suitable abbreviation of the name of any other explosive, if the proof has been made with such, and shall also be stamped with a number indicating the number of grains of this explosive with which the gun may be charged with safety.

Two-bore and other extra large guns have special marks stamped upon them.

The barrels of rifled choke-bore guns are proved with double the quantity of powder, and one and one-half the weight of lead of the service charge (fourth column of adjoined scale).

Herewith are examples of the proof marks of Birmingham and London, as impressed upon shot gun-barrels. The *marks* only, not the figures, will be found also on the breech actions.

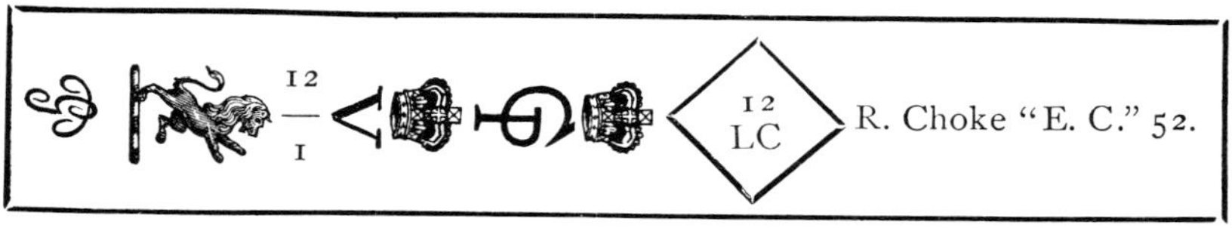

The above marks of the London Proof House signify that the barrel has been twice proved, that the diameter of the barrel is ·740 of an inch, that the chamber is more than three inches long, that the barrel is rifled and choke-bored, and has been proved as required for that description of boring, and that the barrel has also been tested with a granulated gun cotton known as "E.C." gunpowder, of which the standard charge to be used in the gun is 52 grains, or less.

The marks of the Birmingham Proof House, indicating a simple proof of a 12-bore gun.

On the following page will be found the scale of charges used in proving shot guns.

These new rules have not been long enough in force for gunmakers or sportsmen to determine whether or not they will work successfully. It must be admitted that they increase the liability and responsibility of the gunmaker. The gunmaker must declare for what length of cartridge case his gun is intended, and in some instances with what charge of powder it will be used, and the proof will then be made in accordance with his declaration.

SCALE OF PROOF FOR MUZZLE-LOADING SHOT GUNS.

Number of Box.	Diameter of Bore by Calculation.	Provisional Proof. Powder.		Provisional Proof. Bullet.	Definitive Proof. Powder.		Definitive Proof. Shot.		Service Charge. Powder.		Service Charge. Shot.	
		grains	drs.	grains	grains	drs.	grains	oz.	grains	drs.	grains	oz.
4	1·052	928	34	1649	547	20	2041	$4\frac{2}{3}$	273	10	1531	$3\frac{1}{2}$
5	·976	740	27	1315	426	$15\frac{9}{16}$	1623	$3\frac{17}{24}$	213	$7\frac{25}{32}$	1217	$2\frac{25}{32}$
6	·919	612	$22\frac{1}{2}$	1090	357	$13\frac{1}{16}$	1367	$3\frac{1}{8}$	179	$6\frac{17}{32}$	1025	$2\frac{11}{32}$
7	·873	525	$19\frac{1}{4}$	931	308	$11\frac{1}{4}$	1185	$2\frac{17}{24}$	154	$5\frac{5}{8}$	889	$2\frac{1}{32}$
8	·835	481	$17\frac{1}{2}$	812	270	$9\frac{7}{8}$	1057	$2\frac{5}{12}$	135	$4\frac{15}{16}$	793	$1\frac{13}{16}$
9	·803	394	$14\frac{1}{2}$	720	244	$8\frac{15}{16}$	967	$2\frac{5}{24}$	122	$4\frac{15}{32}$	725	$1\frac{21}{32}$
10	·775	372	$13\frac{1}{2}$	646	219	8	875	2	109	4	656	$1\frac{1}{2}$
11	·751	350	$12\frac{3}{4}$	586	191	7	765	$1\frac{3}{4}$	96	$3\frac{1}{2}$	574	$1\frac{5}{16}$
12	·729	350	$12\frac{3}{4}$	535	178	$6\frac{1}{2}$	729	$1\frac{2}{3}$	89	$3\frac{1}{4}$	547	$1\frac{1}{4}$
13	·710	328	12	493								
14	·693	306	$11\frac{1}{4}$	457	164	6	656	$1\frac{1}{2}$	82	3	492	$1\frac{1}{8}$
15	·677	306	$11\frac{1}{4}$	425								
16	·662	295	$10\frac{3}{4}$	399								
17	·649	295	$10\frac{3}{4}$	374	150	$5\frac{1}{2}$	583	$1\frac{1}{3}$	75	$2\frac{3}{4}$	437	1
18	·637	273	10	352								
19	·626	241	$8\frac{3}{4}$	334								
20	·615	219	8	316	136	5	511	$1\frac{1}{6}$	68	$2\frac{1}{2}$	383	$\frac{7}{8}$
21	·605	219	8	300								
22	·596	197	$7\frac{1}{4}$	287								
23	·587	197	$7\frac{1}{4}$	274								
24	·579	186	$6\frac{3}{4}$	262								
25	·571	186	$6\frac{3}{4}$	251								
26	·563	186	$6\frac{3}{4}$	242	109	4	437	1	55	2	328	$\frac{3}{4}$
27	·556	186	$6\frac{3}{4}$	231								
28	·550	186	$6\frac{1}{3}$	223								
29	·543	164	6	214								
30	·537	164	6	207								
31	·531	164	6	200								
32	·526	164	6	194								
33	·520	153	$5\frac{1}{2}$	188								
34	·515	153	$5\frac{1}{2}$	182								
35	·510	153	$5\frac{1}{2}$	177	82	3	328	$\frac{3}{4}$	41	$1\frac{1}{2}$	246	$\frac{9}{16}$
36	·506	153	$5\frac{1}{2}$	172								
37	·501	153	$5\frac{1}{2}$	167								
38	·497	142	$5\frac{1}{4}$	162								
39	·492	142	$5\frac{1}{4}$	158								
40	·488	142	$5\frac{1}{4}$	154								
41	·484	131	$4\frac{3}{4}$	150								
42	·480	131	$4\frac{3}{4}$	146								
43	·476	131	$4\frac{3}{4}$	142								
44	·473	131	$4\frac{3}{4}$	139								
45	·469	120	$4\frac{1}{2}$	136								
46	·466	120	$4\frac{1}{2}$	133	55	2	219	$\frac{1}{2}$	27	1	164	$\frac{3}{8}$
47	·463	120	$4\frac{1}{2}$	130								
48	·459	120	$4\frac{1}{2}$	127								
49	·456	120	$4\frac{1}{2}$	124								
50	·453	120	$4\frac{1}{2}$	122								

SCALE OF PROOF FOR BREECH-LOADING SHOT GUNS.

Number of Box.	Diameter of Bore by Calculation.	Provisional Proof.			Definitive Proof.				Service Charge.			
		Powder.		Bullet.	Powder.		Shot.		Powder.		Shot.	
		grains	drs.	grains	grains	drs.	grains	ozs.	grains	drs.	grains	ozs.
4	1·052	928	34	1649								
5/2	1·026	......	......	......								
5/1	1·001	......	......	......								
5	·976	740	27	1315								
6/2	·957	......	......	......	492	18	1749	4	246	9	1312	3
6/1	·938	......	......	......								
6	·919	612	22½	1090								
7/2	·903	......	......	......								
7/1	·888	......	......	......								
7	·873	525	19¼	931								
8/2	·860	......	......	......								
8/1	·847	......	......	......								
8	·835	481	17½	812	328	12	1167	2⅔	164	6	875	2
9/2	·824	......	......	......								
9/1	·813	......	......	......								
9	·803	394	14½	720								
10/2	·793	......	......	......								
10/1	·784	......	......	......								
10	·775	372	13½	646	219	8	875	2	109	4	656	1½
11/1	·763	......	......	......								
11	·751	350	12¾	586								
12/1	·740	......	......	......								
12	·729	350	12¾	535	178	6½	729	1⅔	89	3¼	547	1¼
13/1	·719	......	......	......								
13	·710	328	12	493								

SCALE OF PROOF FOR BREECH-LOADING SHOT GUNS (*continued*).

Number of Box.	Diameter of Bore by Calculation.	Provisional Proof.			Definitive Proof.				Service Charge.			
		Powder.		Bullet.	Powder.		Shot.		Powder.		Shot.	
		grains	drs.	grains	grains	drs.	grains	ozs.	grains	drs.	grains	ozs.
$\frac{14}{1}$	·701	......	......	......								
14	·693	306	$11\frac{1}{4}$	457	164	6	656	$1\frac{1}{2}$	82	3	492	$1\frac{1}{8}$
$\frac{15}{1}$	·685	......	......	......								
15	·677	306	$11\frac{1}{4}$	425								
$\frac{16}{1}$	·669	......	......	......								
16	·662	295	$10\frac{3}{4}$	399								
$\frac{17}{1}$	·655	......	......	......	150	$5\frac{1}{2}$	583	$1\frac{1}{3}$	75	$2\frac{3}{4}$	437	1
17	·649	295	$10\frac{3}{4}$	374								
18	·637	273	10	352								
19	·626	241	$8\frac{3}{4}$	334								
20	·615	219	8	316								
21	·605	219	8	300	137	5	511	$1\frac{1}{6}$	68	$2\frac{1}{2}$	383	$\frac{7}{8}$
22	·596	197	$7\frac{1}{4}$	287								
23	·587	197	$7\frac{1}{4}$	274								
24	·579	186	$6\frac{3}{4}$	262								
25	·571	186	$6\frac{3}{4}$	251								
26	·563	186	$6\frac{3}{4}$	242	109	4	437	1	55	2	328	$\frac{3}{4}$
27	·556	186	$6\frac{3}{4}$	231								
28	·550	186	$6\frac{3}{4}$	223								
29	·543	164	6	214								
30	·537	164	6	207								
31	·531	164	6	200	82	3	292	$2\frac{2}{3}$	41	$1\frac{1}{2}$	219	$\frac{1}{2}$
32	·526	164	6	194								

Poor as our Government proof test has been, it has ever been the strongest and most thorough of any obligatory test yet imposed in any country. At Liège he proof is done with a quality of powder so poor, that even in the cheapest Continental-made breech-loader both barrels may be proved together, without the probability of a reject. At St.-Étienne, one charge is used for a gun of any calibre, which is absurd. In America and Germany, where no Proof House exists, guns and toys are turned out which are not only a disgrace to their makers, but a positive danger to all who handle them.

The Belgian proof-mark is  but many guns and more barrels made in Belgium, and sold as of English manufacture, bear the London or Birmingham proof-marks.

Another fraud often perpetrated is the imitation of English proof-marks by the very unscrupulous gunmakers abroad, who do not hesitate to impose upon their own countrymen their counterfeit rubbish, which, if sent for sale to England, would at once bring a prosecution upon the vendors. Although nominally there is no difference between the Birmingham and London proof, actually the Birmingham proof has been the more severe.

THE BURSTING STRAIN OF GUN BARRELS.

Thirty years ago the late W. Greener formulated a theory as to the relative strain upon a gun barrel produced by different charges of powder. According to his calculation, if a tube were charged in the ordinary manner, and with 1 oz. of shot, the strain generated on the explosion will be equal to 40,000 lbs., or 1,700 lbs. to the square inch.

Recent experiments with scientifically constructed instruments have proved the above to be pretty nearly correct. To the staff of the *Field* newspaper, led by its indefatigable editor, and ably seconded by the manager of the Schultze Gunpowder Company's manufactory, gunmakers are indebted for valuable technical data, and elaborately formulated conclusions, based upon practically endless experiments, all of which have been of great service to those gunmakers able to make use of them. The art of making shot guns has been elevated to a science, and although full details of the experiments, or even a summary of the figures would be out of place here, the author feels compelled to record the few following facts :—

A 12-bore gun, fired with $2\frac{1}{2}$ drams of powder and one ounce of shot, gives 1,640 lbs. pressure *per square inch* at one inch from the breech, 1,448 lbs. at $2\frac{3}{4}$ inches from the breech, and 916 lbs. at six inches from the breech.

The usual sporting charge of 3 drams and $1\frac{1}{8}$ ozs. gives 2,090 lbs. at 1 inch, 1,796 at $2\frac{3}{4}$ inches, and 1,046 at 6 inches from the breech, whilst the not unusually heavy 12-bore charge of 4 drams and $1\frac{1}{4}$ ozs. gives 3,770 lbs. at 1 inch, 3,210 at $2\frac{3}{4}$ inches, and 1,321 at 6 inches from the breech—more than 70,000 lbs. pressure on the barrel. Equal measure of "Schultze" gunpowder gives *less* pressure upon the barrel at the first distance ; but *more* at the other distances with heavy charges. The charge of 3 drams (42 grains) and $1\frac{1}{8}$ ozs. gives 1,850 lbs. at 1 inch, 1,910 at $2\frac{3}{4}$ inches, and 1,067 at 6 inches. With charges up to and including 3 drams and 1 oz. the Schultze gives less pressure than does black gunpowder, but with heavier charges it gives increased pressure at $2\frac{3}{4}$ and 6 inches ; but with *all* charges gives less pressure at 1 inch from the breech than does the proportionate charge of black.

In guns of smaller bore the proportionate pressure per square inch is greater than that upon 12-bores with the same charge of powder. The charge of $2\frac{1}{2}$ drams and $\frac{7}{8}$ oz. of black in a 20-bore gives 2,825 lbs. at 1 inch, 2,015 at $2\frac{3}{4}$ inches, and 1,080 at 6 inches; with like charge of Schultze, 1,772 at 1 inch, 1,890 at $2\frac{3}{4}$ inches, and 1,046 at 6 inches. Again, with 20-bores, if the charge of Schultze be increased beyond 3 drams and 1 oz. it will give a greater pressure at 6 inches from the breech than will a like charge of black, but a less pressure at 1 inch from the breech. The powder referred to above is Curtis and Harvey's No. 4. The following gives the comparative strengths of the different sizes of powder usually employed in shot guns.

Powder.	At 1 inch.	At $2\frac{3}{4}$ inches.	At 6 inches.
C & H No. 6	1,634 lbs.	1,548 lbs.	980 lbs.
C & H No. 4	2,090 ,,	1,796 ,,	1,046 ,,
C & H Fine Basket	2,900 ,,	2,097 ,,	1,111 ,,
C & H No. 4	3,430 ,,	2,385 ,,	1,090 ,,
Schultze	1,850 ,,	1,910 ,,	1,067 ,,

These figures were all obtained under exactly the same conditions, and are the average of many shots, and may therefore be accepted as a standard with the 12-bore with same charge.

There are several things which will cause an increase on the pressure exerted by any one charge in a barrel of any bore; one of these is the cap. The figures given above were registered with Eley's caps in Eley's ordinary cases. Many sportsmen now use the "Life" cases, which have a larger and stronger cap than Eley's, and which cap affects the pressure with Schultze powder very considerably; for instance, a 12-bore with 3 drams and $1\frac{1}{8}$ ozs. of shot gives in Eley's case a pressure of 1,850 to 1,910 lbs., with the large cap from 1,810 to 1,895 lbs. per square inch; with $3\frac{1}{2}$ drams and $1\frac{1}{4}$ ozs., with Eley's cap 2,440 lbs. to 2,680 lbs., with the large cap 2,730 to 2,975 lbs. And in the 20-bore with $2\frac{1}{2}$ drams and $\frac{7}{8}$-oz., with Eley's cap 1,972 lbs. to 2,230 lbs., and with the large cap 3,060 lbs. to 3,210 lbs.

Another and much more important cause of increased pressure is the presence of an obstruction in the barrel, and this not unfrequently results in a burst. It will be as well, in addition to the actual figures, to refer to other facts.

The figures were obtained by the same method as that employed for registering the pressure upon the barrel with various charges and explosives already referred to; the obstruction consisting of a tight "Field" wad, thick felt and card wad, $1\frac{1}{8}$ ozs. of No. 6 Chilled Shot, and card wad over the shot, measurements taken from the breech to the "Field" wad. Charge used 3 drams, and $1\frac{1}{8}$ ozs. of No. 6 shot.

	At 1 inch.	At $2\frac{3}{4}$ inches.	At 6 inches.
Without any obstruction...	1,835.........	1,907.........	1,058 lbs. pressure.
Obstruction at 24 in.	1,820.........	1,907.........	1,076 ,, ,,
,, 18 in.	1,832.........	1,900.........	1,114 ,, ,,
,, 12 in.	1,805.........	1,918.........	1,132 ,, ,,
,, 9 in.	1,826.........	1,930.........	1,147 ,, ,,
,, 7 in.	1,835.........	1,977.........	1,290 ,, ,,
,, 6 in.	1,850.........	2,030.........	1,266 ,, ,,

The pressure at the obstruction has not yet been ascertained. The facts are that some obstructions will invariably burst some barrels, whilst other obstructions are not such as to produce a bursting pressure.

A felt wad $\frac{3}{8}$-inch thick, and fitting the barrel tightly, if placed at 6 inches from the breech, will not cause the barrel to burst when the next charge is fired. Neither will such obstruction as shot wads, cobwebs, or leaves, cause a barrel to burst; a barrel, however, plugged at the muzzle with *mud*, will, if fired with the obstruction in, burst at the muzzle, and snow *may* sometimes effect the same result, but the author has never found it to do so.

A charge of shot which shall slip to the muzzle, or be kept between the wads in the barrel at a distance from the charge, *may* and indeed probably will cause the barrel to burst if the charge of powder, also loose, be fired, and it will certainly produce a burst if a charge of shot be fired through the barrel having this obstruction.

Owing to defective cartridges, it sometimes happens that the charge is not blown from the muzzle, only to it, and sportsmen—not having noticed the very weak discharge—again load and fire.

The looser the shot, *i.e.*, the greater the distance between the two wads, the worse the burst.

Felt wads placed all down the barrel commencing at six inches from the chamber and then at intervals of two inches, will in a good barrel be blown out by the charge, powder and shot, without injuring the barrel.

The steel barrel of same weight and size as one of best Damascus or other hand-forged barrel of best material will burst with a less strain, but as to what pressure each will bear has not yet been accurately ascertained; but although a best steel barrel may be broken or burst with a charge of $21\frac{1}{4}$ drams, it will require 23 drams to burst a twist barrel of the same weight and proportions.

Velocity is the test of "shooting." Velocity generally means a good pattern, and a good pattern at a long range is convertible with velocity; penetration is velocity. The velocity of shot depends upon the nature and quality of the explosive used, the perfect finish and shape of the barrel, the rotundity, size, and specific gravity of the pellets. The details of the various deviations in results obtainable by different grains of powder, different gauges of guns, different sizes of shot, chokes and cylinders, would be tedious. Again must the author ask the reader's indulgence, and give only such averages as may be of greatest importance in helping the sportsman to arrive at a decision as to what gauge and charge will best suit his requirements.

The choke-bore shoots more regularly than does the cylinder, and sends its pellets more compactly; and by compactly is meant not only that the divergence from the centre, as shown on a plate, is less, but that there is much less space between the first and last pellet when the body of the pellets shall reach a certain point. Hence it is that in computing the muzzle velocity of the two guns calculated upon the time occupied by the *first* pellet in travelling 120 feet, the choke shows to a disadvantage. The author, however, from his experiments, would give the choke an average velocity of 5 feet per second more than the cylinder when using black gunpowder, and 25 feet per second more when using nitro-compound.

Of the difference in the velocities of gunpowders, black gunpowder has not the velocity of nitro-compounds, being inferior to the extent of about 20 feet with standard charge.

As to cartridge cases, the "Life" cases, with large caps, give higher velocities generally than do Eley's cases, whilst brass cases in guns specially constructed for them give higher velocities than paper cases with black powder, but lower with nitro-compounds than the "Life" cases charged with the same powder.

The bore of the gun affects velocity as follows :—

20-BORE GUN, with $2\frac{1}{4}$ drams and 1 oz. No. 6 shot, average velocity, 725 feet; with same charge, but No. 5 shot, average velocity, 738·8 ft. per second.

16-BORE GUN, with $2\frac{1}{2}$ drams, and 1 oz. No. 6 shot, average velocity, 780 ft.; with same charge, but No. 5 shot, 791 ft.

12-BORE GUN, with $3\frac{1}{2}$ drams and $1\frac{1}{8}$ ozs. No. 6. shot, average velocity, 842·171 ft.

10-BORE GUN, with $4\frac{1}{2}$ drams and $1\frac{1}{2}$ ozs. No. 6 shot, average velocity, 890 ft.; with same charge, but No. 4 shot, 936 ft.; with $1\frac{1}{2}$ ozs. No. 1 shot and 5 drams of powder, 943 ft.

8-BORE GUN, with 6 drams of powder, paper case, and $2\frac{1}{4}$ ozs. No. 1 shot, average velocity, 907 ft.; with 7 drams No. 4 powder and $2\frac{1}{4}$ ozs. No. 1 shot, and brass case, average velocity, 984 feet.; with same load, but finer-grained powder, 945 ft.; with same load, but with ducking powder, *expressly manufactured for* 8- *and* 4- *bore duck guns,* average velocity *only* 904 ft.

With 3 drams and 1 oz. of No. 6 the muzzle velocities should be with 12-bore about 870 feet per second; with 16-bore, about 885 feet per second; with 20-bore, about 920 feet per second.

The size of the shot is an important factor in computing velocity, as it is in fixing the killing range of the gun or judging its penetration. The figures above given are the calculated muzzle velocities; but the following

F 2

summary gives the actual mean velocity in feet per second of the body of the charge of shot at the range indicated, measured by Mr. R. W. S. Griffith :—

CHARGE.	5 Yds.	10 Yds.	15 Yds.	20 Yds.	25 Yds.	30 Yds.	35 Yds.	40 Yds.	45 Yds.	50 Yds.	55 Yds.	60 Yds.
2½ drms. 1 oz. No. 1	1039	1022	1001	979	951	929	904	880	856	829	802	780
3 ,, 1 ,, ,, 1	1185	1168	1150	1120	1076	1039	992	939	919	880	852	831
3 ,, 1⅛ ,, ,, 1	1169	1140	1126	1089	1054	1003	962	935	914	891	861	825
3½ ,, 1⅛ ,, ,, 1	1220	1198	1175	1143	1103	1060	1012	966	938	914	890	876
3½ ,, 1¼ ,, ,, 1	1172	1151	1130	1111	1073	1029	989	949	908	867	835	799
4 ,, 1¼ ,, ,, 1	1239	1221	1205	1181	1144	1106	1066	1022	976	936	909	863
2½ ,, 1 ,, ,, 5	996	975	953	924	886	850	831	792	771	744	710	684
3 ,, 1 ,, ,, 5	1160	1133	1106	1066	1021	969	922	879	840	801	764	729
3 ,, 1⅛ ,, ,, 5	1127	1034	1070	1047	1014	970	914	875	835	790	741	672
3½ ,, 1⅛ ,, ,, 5	1182	1164	1136	1088	1045	1002	960	907	875	838	799	757
3½ ,, 1¼ ,, ,, 5	1130	1111	1088	1058	1026	979	931	880	834	790	741	689
4 ,, 1¼ ,, ,, 5	1207	8190	1171	1134	1094	1051	994	932	892	853	810	764
2½ ,, 1 ,, ,, 6	990	963	941	905	863	829	804	776	752	722	690	651
3 ,, 1 ,, ,, 6	1154	1130	1100	1061	1012	950	904	862	826	770	730	694
3 ,, 1⅛ ,, ,, 6	1119	1091	1063	1035	999	942	890	850	808	769	717	652
3½ ,, 1⅛ ,, ,, 6	1175	1153	1120	1079	1032	986	940	894	861	825	780	723
3½ ,, 1¼ ,, ,, 6	1121	1100	1081	1050	1015	970	912	858	810	762	714	663
4 ,, 1¼ ,, ,, 6	1199	1177	1159	1122	1082	1034	977	907	850	816	774	734
2½ ,, 1 ,, ,, 10	922	892	872	841	802	760	703	670	622	551	465	370
3 ,, 1 ,, ,, 10	1120	1094	1071	1029	959	891	809	751	704	630	559	440
3 ,, 1⅛ ,, ,, 10	1076	1041	999	941	886	831	775	710	540	465	430	375
3½ ,, 1⅛ ,, ,, 10	1126	1096	1060	1012	965	892	830	781	711	630	550	460
3½ ,, 1¼ ,, ,, 10	1066	1041	1017	980	940	896	834	760	706	620	540	446
4 ,, 1¼ ,, ,, 10	1145	1115	1090	1045	987	927	964	792	720	642	551	495

All these figures were obtained with a Greener-choked 12-bore gun, with Schultze powder and good wadding. Inferior wadding will produce lower velocity, especially at the longer ranges.

With an 8-bore gun, fired with 6½ drams of black powder, No. 4, or the equivalent of Schultze, and 2 ozs. of shot, the velocities will average as under :—

SIZE OF SHOT.	VELOCITY OBTAINED AT		
	60 Yds.	70 Yds.	80 Yds.
No. 1	892	828	748
,, 2	872	804	718
,, 3	850	775	686
,, 4	825	744	651
,, 5	798	710	612
,, 6	768	672	566

THE STRINGING OF A CHARGE OF SHOT.

That the pattern of a gun, as seen upon a target, does not reveal the manner in which the pellets arrive, is a well-known fact, and the gear necessary to exhibit more precisely the manner of the flight of the shot is of such a cumbrous and complex nature, that it is doubtful if it will ever come into general use or be easily understood.

As already stated, the gun which shoots best must make the closest pattern, and a pattern which reveals the least deviation of the pellets from the common centre is also a proof that individual pellets have been less in advance, and in rear, of the main body of shot during flight than would have been the case had the spread upon the target been larger.

From experiments made it has been proved that the 12-bore gun with 42 grains of Schultze powder, and 304 pellets of No. 6 shot, gives most regular patterns, and that the regularity of the shooting is enhanced by the barrel being properly choke-bored. It has also been proved that a larger charge of powder, although it increases the velocity of the shot, impairs the regularity of the pattern. Further experiments have proved that not only does the choke-bore send its shot closer and more regularly, but also more compactly. To what extent shot "strings," will be gathered from the following figures and diagrams:—With a cylinder gun, with 42 grains of "Schultze," and 304 pellets of No. 6 shot, the first pellets reach the target at 40 yards' distance in ·138 second, whereas the last pellets do not reach it until ·187 second; consequently, whilst the first pellets may strike a bird at 40 yards, the slower pellets have not reached a distance of 30 yards from the muzzle of the gun.

As a matter of fact, about five per cent. of the pellets of the charge arrive simultaneously at the target at 40 yards' distance from the gun; these pellets are *closely* followed by 25 per cent. to 30 per cent. of the pellets of the charge if the gun be a good one, and this 30 per cent. to 40 per cent. of the pellets represents practically the actual killing value of the shot, for the remaining pellets flying irregularly, and at a much lower velocity, tail off so rapidly that little reliance can be placed upon them. The fac-simile targets shown exhibit the usual pattern faithfully, being a photographic reduction of the actual diagrams; but to show accurately, and on the same scale, the side view illustrating the pellets in flight at 60 yards from a cylinder gun, would require a diagram nearly five feet in length. The annexed diagrams and table will enable those interested to calculate the approximate distances between the pellets of the charge at any distance.

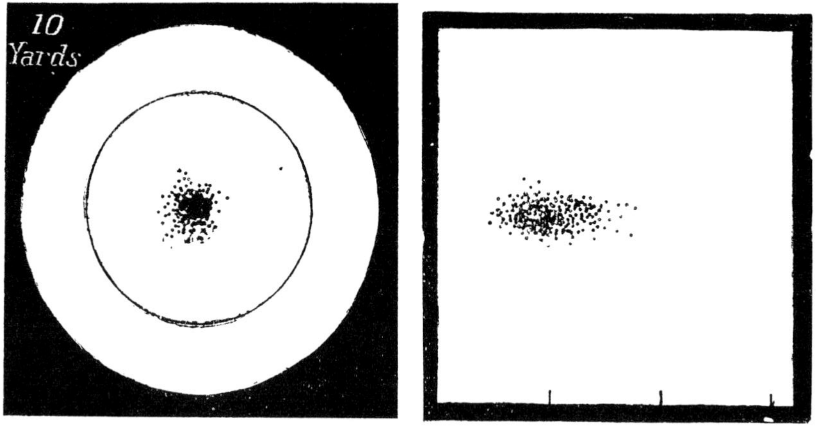

Obtained with a Choke-bore Gun loaded with 42 grains of Schultze Gunpowder and $1\frac{1}{8}$ oz. No. 6 Chilled Shot (304 pellets).

FIXED TARGET (4 FT. DIAMETER).

MOVING TARGET ($\frac{1}{26}$ SCALE).

DISTRIBUTION OF THE 304 PELLETS.

At 10 yards.—All in the 30-in. circle.

At 20 yards.—Ditto.

At 30 yards.—**278** in the 30-in. circle; **24** in the 30–48-in. belt; and **2** outside the 4-ft. circle.

At 40 yards.—**233** in the 30-in. circle; 65 in the 30–48-in. belt; and 6 outside the 4-ft. circle.

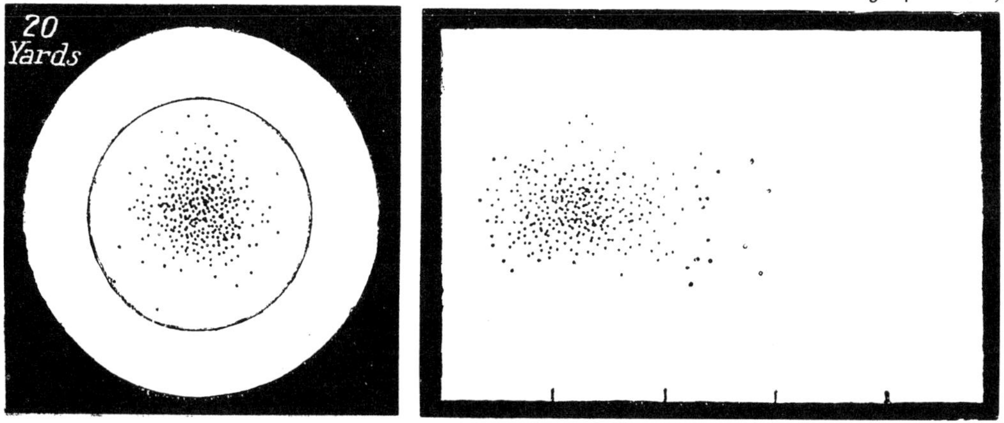

At 50 yards.—160 in the 30-in. circle; 90 in the 30–48-in. belt; and **54** outside the 4-ft. circle.

At 60 yards.—100 in the 30-in. circle; 95 in the 30–48-in. belt; and 109 outside the 4-ft. circle.

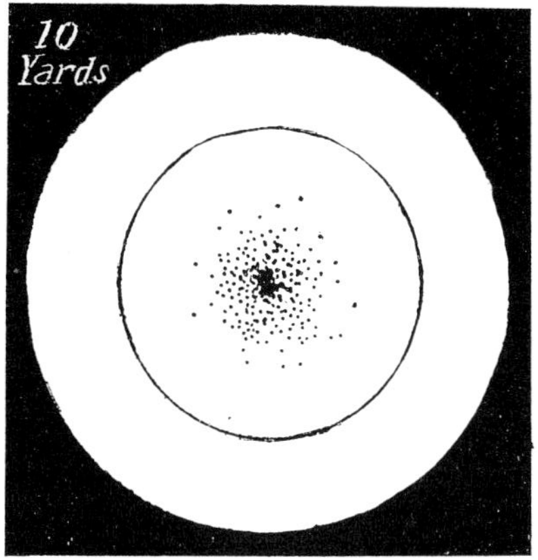

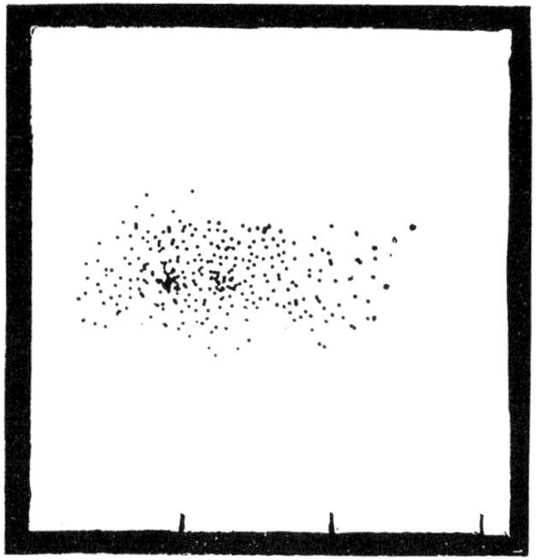

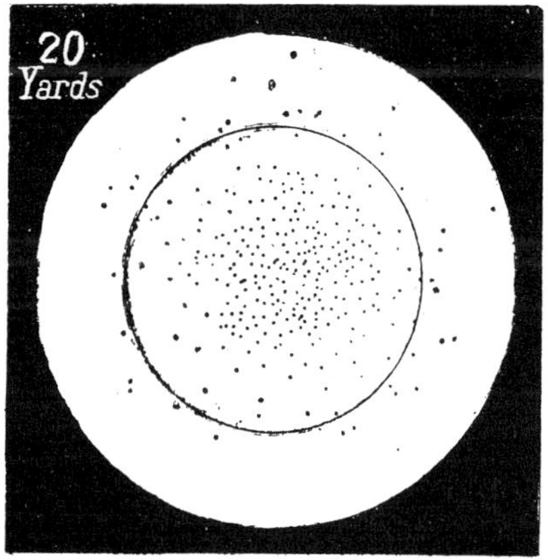

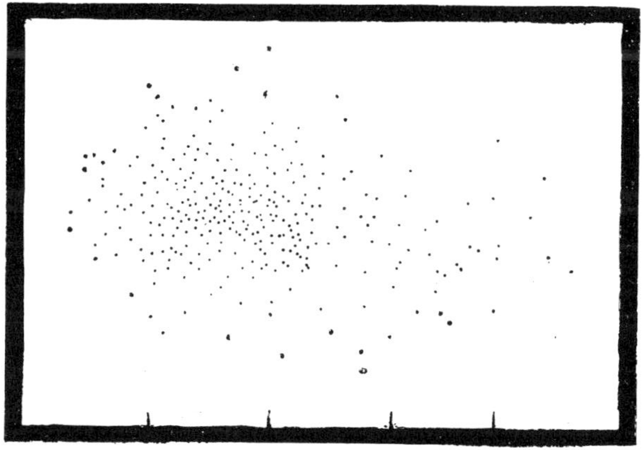

Obtained with a Cylinder Gun loaded with 42 grains of Schultze Gunpowder and $1\frac{1}{8}$ oz. of No. 6 Chilled Shot (304 pellets).

FIXED TARGET (4 FT. DIAMETER).
MOVING TARGET ($\frac{1}{25}$ SCALE).

DISTRIBUTION OF THE 304 PELLETS.

At 10 yards.—All in the 30-in. circle.

At 20 yards.—264 in the 30-in. circle; 38 in the 30–48-in. belt; and 2 outside the 4-ft. circle.

At 30 yards.—172 in the 30-in. circle; 90 in the 30–48-in. belt; and 42 outside the 4-ft. circle.

At 40 yards.—130 in the 30-in. circle; 103 in the 30–48-in. belt; and 71 outside the 4-ft. circle.

At 50 yards.—76 in the 30-in. circle; 86 in the 30–48-in. belt; and 142 outside the 4-ft. circle.

At 60 yards.—61 in the 30-in. circle; 57 in the 30–48-in. belt; and 186 outside the 4-ft. circle.

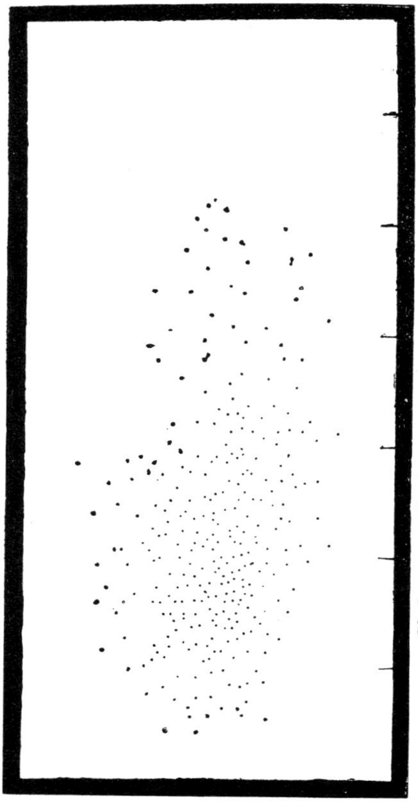

Choke at 30 Yards.

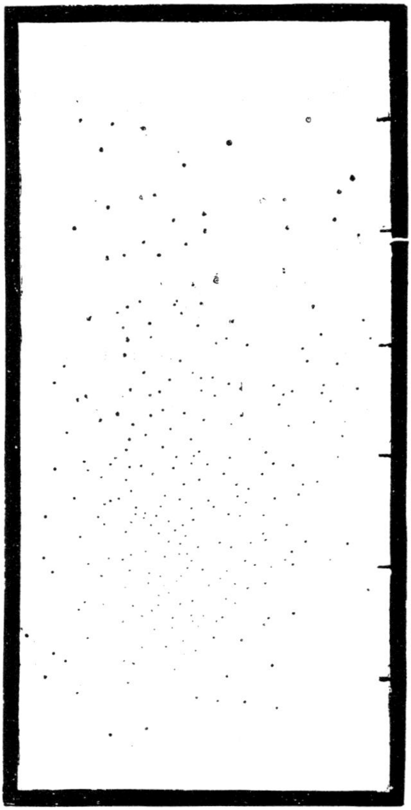

Cylinder at 30 Yards.

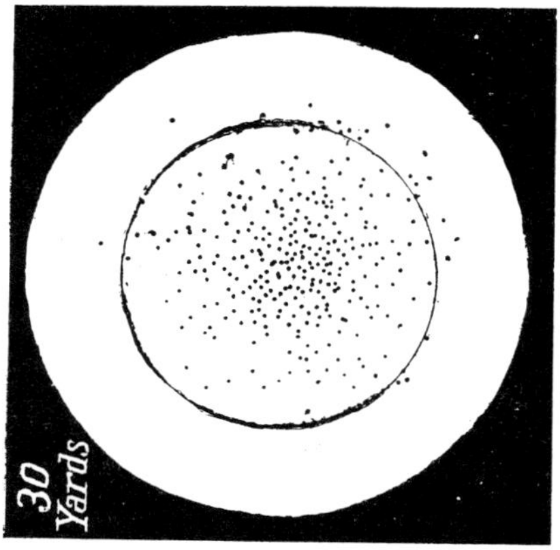

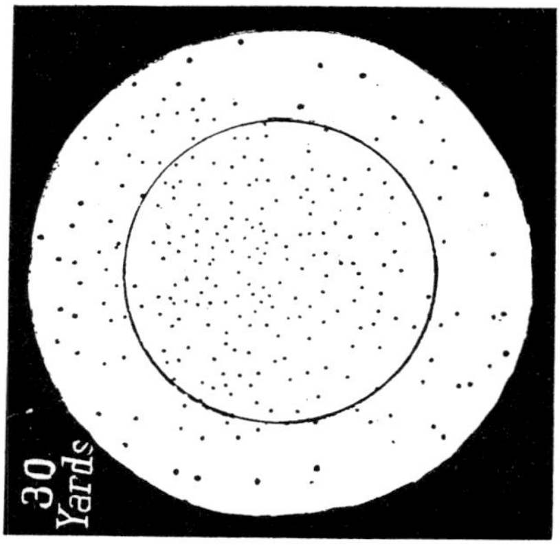

Choke at 40 Yards.

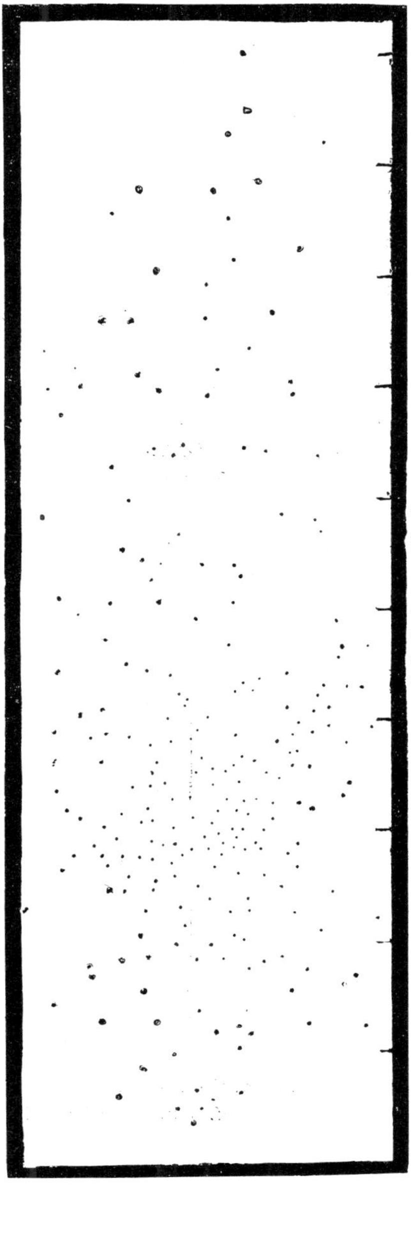

Cylinder at 40 Yards.

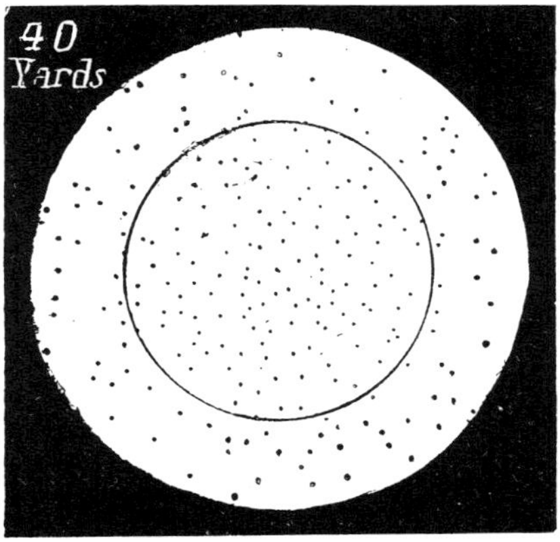

Pattern of the Cylinder Gun at 40 Yards.

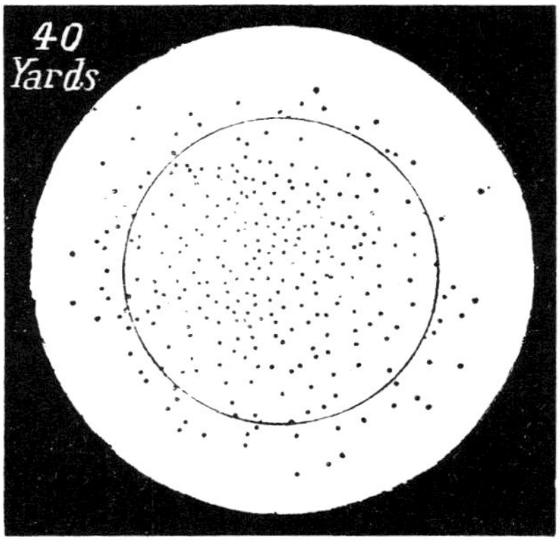

Pattern of CHOKE-BORED Gun at 40 Yards.

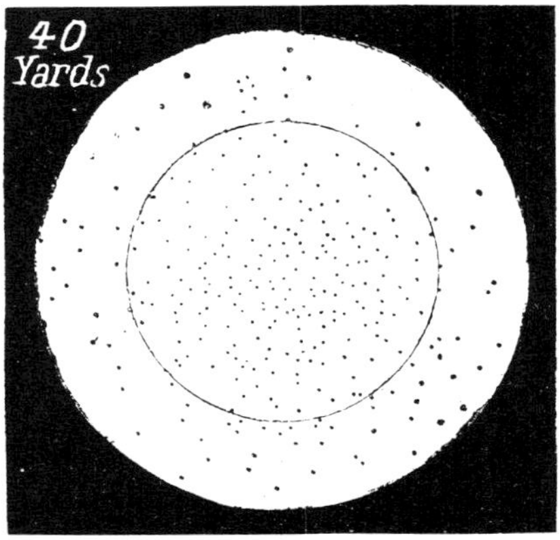

Pattern of the same CHOKE-BORED Gun at 40 Yards, with 49 grains of Schultze Powder and $1\frac{1}{8}$ oz. No. 6 Chilled Shot.

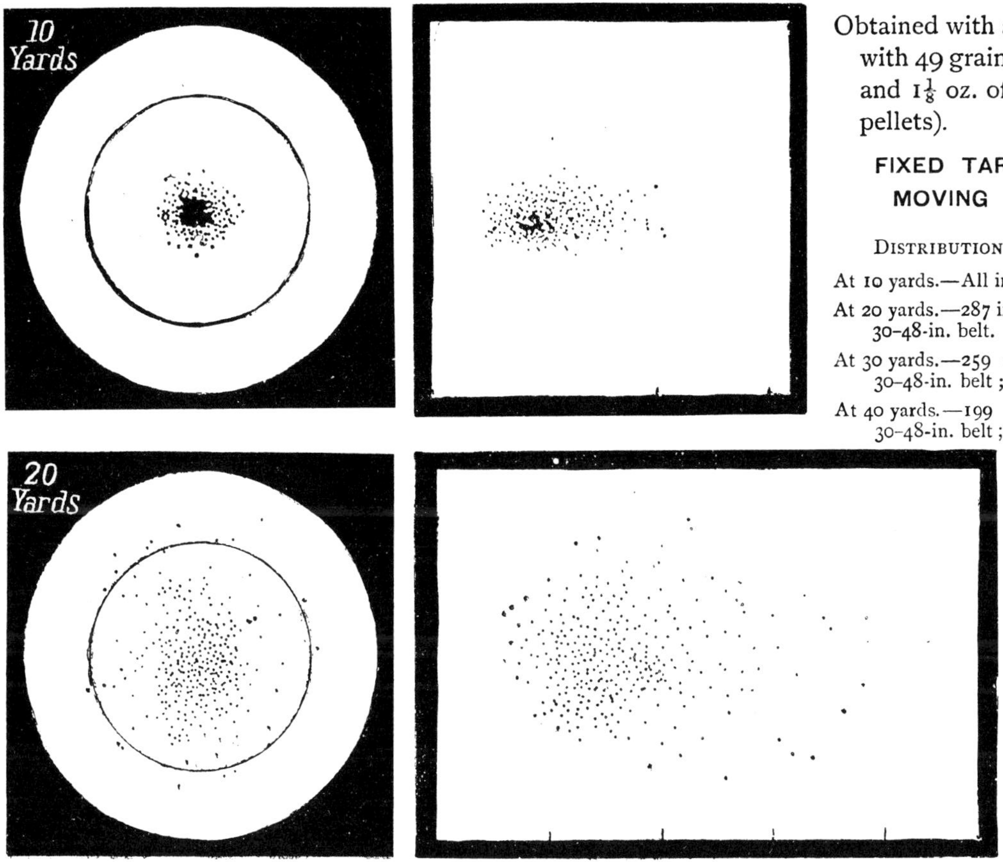

Obtained with a Choke-bore Gun loaded with 49 grains of Schultze Gunpowder and $1\frac{1}{8}$ oz. of No. 6 Chilled Shot (304 pellets).

FIXED TARGET (4 FT. DIAMETER).

MOVING TARGET ($\frac{1}{25}$ SCALE).

DISTRIBUTION OF THE 304 PELLETS.

At 10 yards.—All in the 30-in. circle.

At 20 yards.—287 in the 30-in. circle; and 17 in the 30–48-in. belt.

At 30 yards.—259 in the 30-in. circle; 40 in the 30–48-in. belt; and 5 outside the 4-ft. circle.

At 40 yards.—199 in the 30-in. circle; 89 in the 30–48-in. belt; and 16 outside the 4-ft. circle.

At 50 yards.—135 in the 30-in. circle; 108 in the 30–48-in. belt; and 61 outside the 4-ft. circle.

At 60 yards.—93 in the 30-in. circle; 92 in the 30–48-in. belt; and 119 outside the 4-ft. circle.

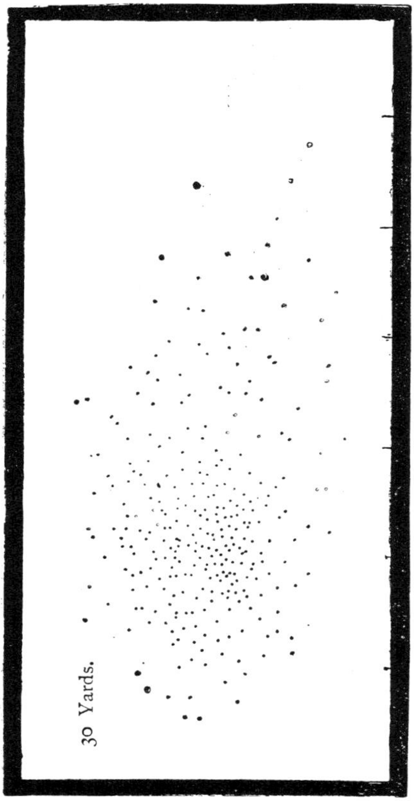
30 Yards.

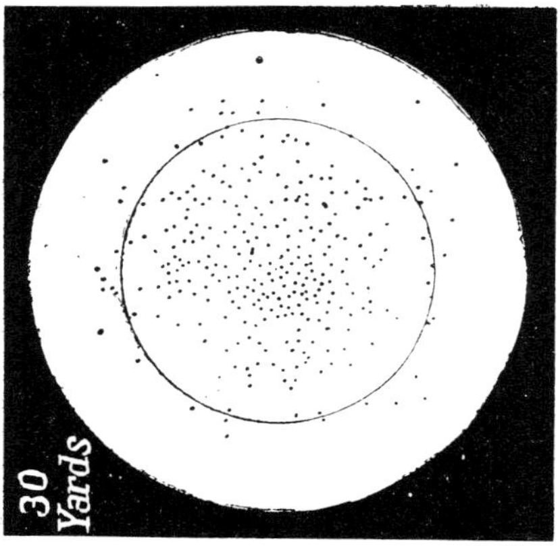
30 Yards

40 Yards.
Last 10 Pellets
One half pattern
25% of pattern =72
C 20% of pattern
Secd
.0062
.0077
.0130
.0425
Velocity to
936 feet sec
850 ft
per second at these centres

Table Showing the Various Velocities Attained by the Pellets of a Charge.

Gun and Load.	Range yards.	First cluster of pellets reach the target.		25 pr cent. pattern lag behind.	Total time for the range.	Equivalent velocity.
		Chronograph velocity.	Equivalent time.			
		Ft.-Sec.	Sec.	Sec.	Sec.	Ft.-Sec.
First Series.	10	1132	·0265	·0010	·0275	1091
W. W. Greener Choke-bore, with 1⅛oz. No. 6	20	1073	·0559	·0020	·0579	1036
shot, and 42 grains of	30	976	·0923	·0035	·0958	939
"Schultze" powder.	40	884	·1357	·0055	·1412	850
	50	802	·1871	·0077	·1948	770
	60	675	·2668	·0102	·2770	650
Second Series.						
W. W. Greener Choke-bore, with 1⅛oz. No. 6	10	1209	·0248	·0012	·0260	1153
shot, and 49 grains of	20	1124	·0534	·0022	·0556	1079
"Schultze" powder.	30	1031	·0873	·0040	·0913	986
* This one is marked on	40*	938	·1280	·0062	·1342	894
photograph to show	50	855	·1754	·0090	·1844	814
method of calculating.	60	756	·2381	·0112	·2493	722
Third Series.	10	1128	·0266	·0015	·0281	1067
W. W. Greener Cylinder, with 1⅛oz. No.	20	1062	·0565	·0030	·0595	1008
6, and 42 grains of	30	963	·0935	·0050	·0985	914
"Schultze" powder.	40	870	·1380	·0080	·1460	823
	50	784	·1915	·0102	·2017	743
	60	667	·2700	·0120	·2820	638

The adjoining diagrams will give at a glance an approximate idea of the difference in the flight of a charge of shot from a choke-bore and a cylinder gun, and also the difference caused by an increased charge of powder in the choke, but as the velocity varies at the different ranges, the diagrams do not show accurately the approximate divergence at all ranges. On the 40-yard diagram one inch is equal to eight feet horizontally, and to two feet only measured perpendicularly. It should be borne in mind that these diagrams were made with an acknowledged good gun, and with cartridges most carefully loaded, by Mr. R. W. S. Griffith, of the "Schultze Powder Co.," for certain experiments, the results of which, with the diagrams here reproduced, and several others, will be found in the 69th volume of the *Field.*

A few further particulars respecting the flight of a charge of shot may be of use to the sportsman. With the usual charge of 3 drams to $1\frac{1}{8}$ ozs. of No. 6, the spread of the charge at 5 yards from the muzzle of a choke-bored gun will be about 5 inches, at 10 yards about 8, at 15 yards 12 inches; with No. 2 shot the spread will be about $1\frac{3}{4}$ inches less at each range; and with No. 8 shot will be very little more than with No. 6 at 5 yards, but $2\frac{1}{2}$ inches more at 10 yards, and 4 inches more at 15 yards. If the charge of *powder* is increased, the spread of the shot at these ranges is increased. In a 12-bore gun charges of more than $3\frac{1}{4}$ drams do not generally give greater penetration to the majority of the pellets, although a *few* pellets of the charge have a greater velocity. No. 6 shot, having a velocity of 500 feet per second, should penetrate 18 sheets of a "Pettit" pad, and will be equal to an energy of 0·90 foot-pounds. No. 3 shot at the same velocity should penetrate 23 sheets, and will equal 1·76 foot-pounds; whilst No. 8 shot at same velocity will penetrate but 16 sheets, and have an energy equal to 0·56 foot-pounds. A velocity of 700 feet is equal to a penetration of 36 sheets with No. 6, of 39 sheets with No. 5, of 47 sheets with No. 2, of 31 sheets with No. 8.

RECOIL is dependent upon the boring of the gun, and will vary with the charge and the nature of the explosive used, the quantity and size of the shot. The recoil should never exceed 100 lbs., and the recoil of a light 20-bore, which should be equal in lbs. to that from an ordinary 12-gauge, would be much more appreciable. With a 12-bore gun weighing 7 lbs. the average recoil with standard charge should not be more than 100 lbs. The difference in the charge will affect it somewhat as follows:—

Charge—	$2\frac{1}{2}$ drs., $\frac{7}{8}$ oz.	$2\frac{1}{2}$ drs., 1 oz.	3 drs., $1\frac{1}{8}$ oz.	3 drs., $1\frac{1}{4}$ oz.	3 drs., $1\frac{1}{2}$ oz.
Recoil—	80	90	100	106	115

The sizes of shot as follows—3 drams and $1\frac{1}{4}$ oz.:—

Shot	No. 2	3	4	5	6	7	8
Recoil	89	90	90	100	106	106	106

The difference in the grain and nature of the explosive as follows—charge 3 drams and $1\frac{1}{8}$ oz., No. 6:—

E.C.	Schultze.	No. 3 grain Black.	No. 4 grain Black.	No. 6 grain Black.
87	86	109	100	85

Wadding and turning over the cartridge case, with pressure, will also cause the recoil to vary slightly, but not sufficiently to require details.

PATTERNS.

The pattern is the *shown* shooting of a gun, the only visible proof of a gun's powers—the killing of game being in some measure dependent upon the skill of the sportsman. The *pattern* of a gun, besides being the most reliable test applicable to a gun, is fortunately the easiest. To ascertain a gun's merits, roughly fire it at the largest sheet of paper obtainable—for most guns 40 yards will be found the best distance. For comparative results count the number of perforations formed in a circle 30 inches in diameter marked upon the paper. A good close pattern is a guarantee that the gun has sufficient force to kill at that distance. The greater the velocity of the *mass* of pellets of the charge the closer is the pattern. No close-shooting gun has inferior penetration, and, generally speaking, the less distant each individual pellet is from the common centre depicted on the target, the less is the distance between the first and last pellets of the charge—in other words, a *close* pattern means uniform velocity in the pellets of the charge. Occasional bad patterns, or patchy patterns, prove the gun to be improperly bored. The closer the pattern at 40 yards, the longer the killing range of the gun.

There have been eight great gun trials, and at all of these—except that of 1879—the pattern has been an important factor in determining the figure of merit.

TABLES OF THE PATTERNS MADE BY THE BEST GUNS IN THE LONDON GUN TRIALS OF 1859, 1866, 1875, 1878, 1879, AND THE AMERICAN GUN TRIALS OF 1873, 1874, 1879.

		drams. oz. shot.		Pattern. Right. Left.
LONDON GUN TRIAL, 1859—				
Muzzle-loader	12-bore	$2\frac{3}{4} \times 1\frac{1}{4}$ No. 6	290 pellets to oz.	158 118
Breech-loader	12-bore	$3 \times 1\frac{1}{4}$ No. 6		144 90
LONDON GUN TRIAL, 1866—				
Breech-loader	12-bore	$2 \times 1\frac{1}{8}$ No. 6	280 pellets to oz.	131 123
Breech-loader	16-bore	$2\frac{1}{2} \times 1$ No. 5		100 118
LONDON GUN TRIAL, 1875—				Average.
*Breech-loader	12-bore	$3\frac{1}{4} \times 1\frac{1}{8}$ No. 6	270 pellets to oz.	214
*Breech-loader	10-bore	$4 \times 1\frac{1}{2}$ No. 6		241
*Breech-loader	20-bore	$2\frac{1}{8} \times 1$ No. 6		145
*Breech-loader	8-bore	$6 \times 2\frac{1}{2}$ No. 6		358
LONDON GUN TRIAL OF EXPLOSIVES, 1878—				
Breech-loader	12-bore	$3\frac{1}{4} \times 1\frac{1}{8}$ No. 6		220·08
LONDON GUN TRIAL, 1879—			Chilled shot.	
Breech-loader	12-bore	$3\frac{1}{4} \times 1\frac{1}{8}$ No. 6	270 pellets to oz.	223
Breech-loader	16-bore	$2\frac{3}{4} \times 1$ No. 6		174
Breech-loader	20-bore	$2\frac{1}{4} \times \frac{7}{8}$ No. 6		174
NEW YORK GUN TRIAL, 1873—				
Breech-loader	12-bore	$3 \times 1\frac{1}{8}$ No. 6	Shot with paper shell	150
Breech-loader	12-bore	$3\frac{1}{2} \times 1\frac{1}{8}$ No. 6	Shot with metal shell	211
Breech-loader	10-bore	$4\frac{1}{2} \times 1\frac{1}{4}$ No. 6	Shot with paper shell	211
CHICAGO GUN TRIAL, 1874—				
Breech-loader	12-bore	4×1 No. 7	309 pellets to oz.	180
Breech-loader	10-bore	$4\frac{1}{2} \times 1$ No. 7		191
CHICAGO GUN TRIAL, 1879—				
Breech-loader	12-bore	$3\frac{1}{2} \times 1\frac{1}{8}$ —	291 pellets to oz.	170
Breech-loader	10-bore	$4 \times 1\frac{1}{4}$ —		200
Breech-loader	16-bore	$2\frac{1}{2} \times 1$ —		163
Breech-loader	20-bore	$2\frac{1}{2} \times 1$ —		138

* These four guns were shot in the selected circle, and with chilled shot.

It need hardly be added that in no case was any partiality shown, nor was there cause for it.

At most of these public gun trials the author's guns have beaten all comers, yet the guns with which he competed were not by any means exceptionally good. Many that he and others have since made would record much better results than those obtained at these gun trials. To

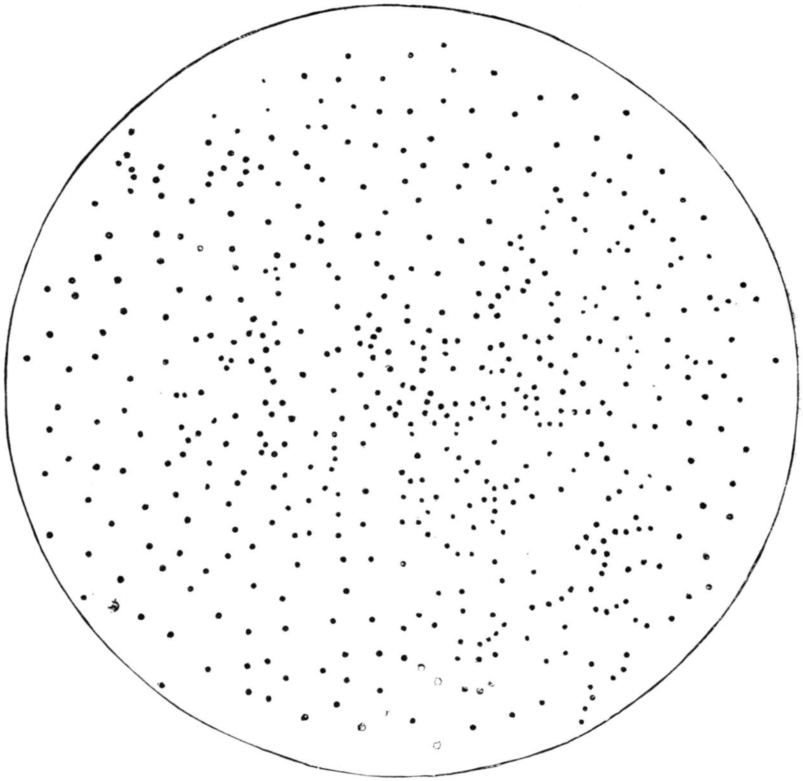

The Winning Pattern at the Leavenworth Trial, made with a W. W. Greener Gun and "Quick Shot" Powder.

cite an instance: the gun made by the author for the Schultze Gunpowder Company has in their hands scored remarkably well, and very many important trials made by them have been made public.

The annual performance of this gun is given under the heading of "Choke-boring;" to this may be added a fac-simile pattern (reduced by photography) from a target 30 inches in diameter, made with W. W. Greener's Patent Treble Wedge-Fast Gun, No. 19,304, built in 1878, specially for the use of the Schultze Gunpowder Company, Limited. Over 100,000 shots were fired from this gun. This is a fac-simile of pattern made by the Schultze Company in 1885, without any alteration to the gun. The gun is admitted to be as good as new in every way.

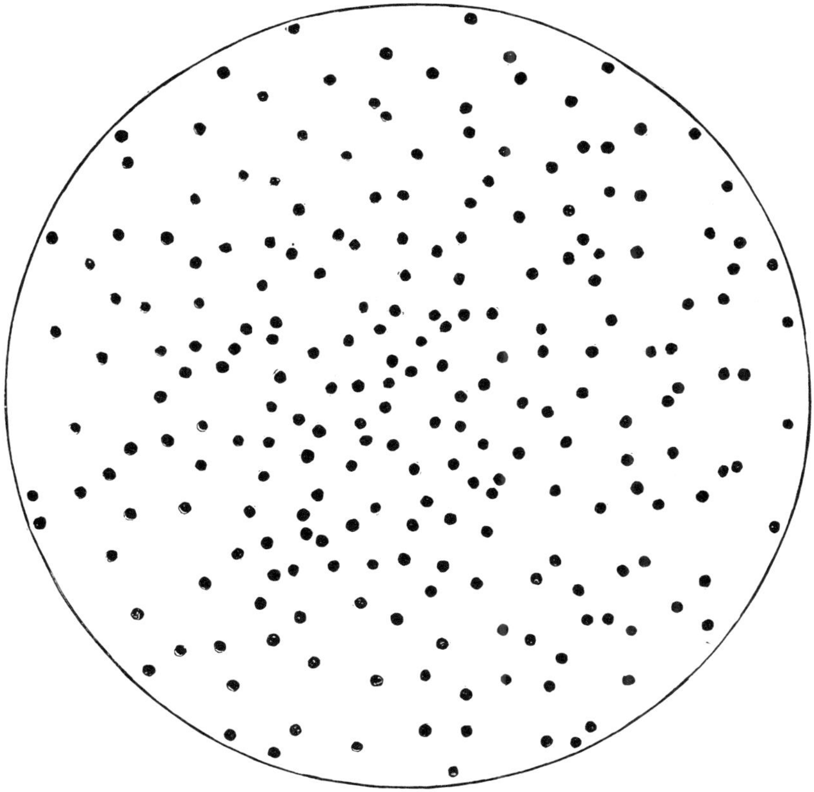

Fac-simile of the Shooting of a W. W. Greener Gun with Schultze Powder and No. 6 Shot.

G

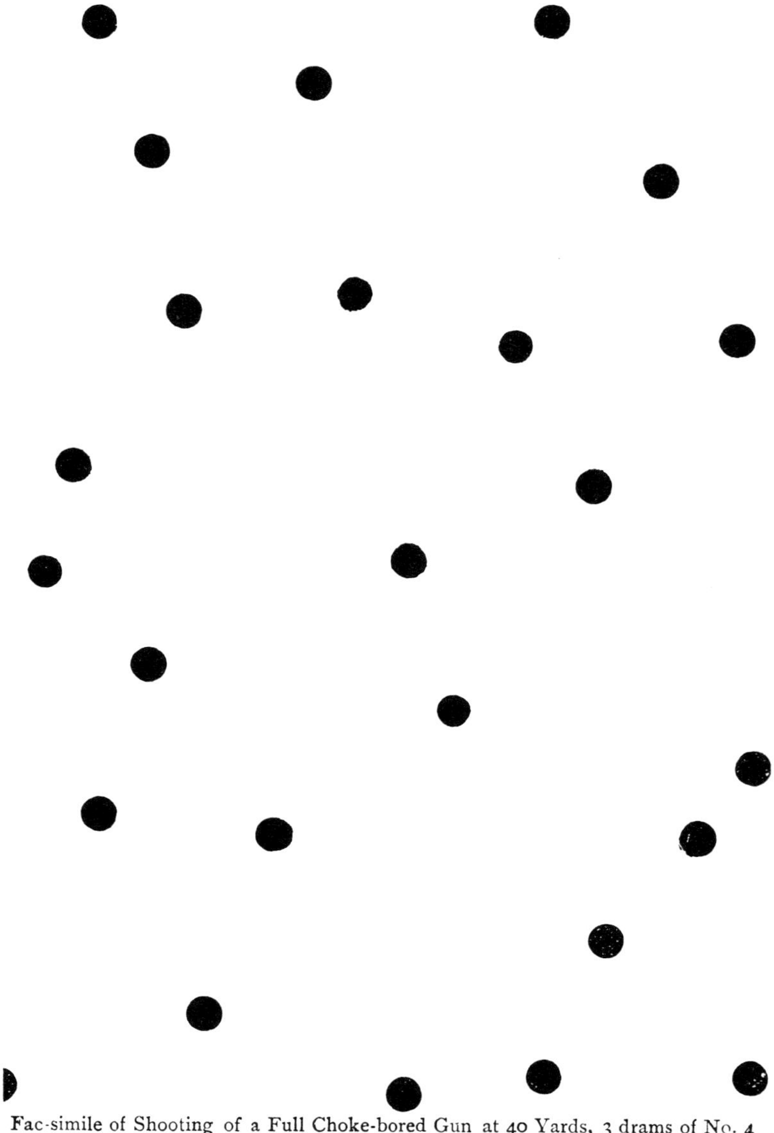

Fac-simile of Shooting of a Full Choke-bored Gun at 40 Yards, 3 drams of No. 4 Powder, and $1\frac{1}{8}$ oz. No. 6 Shot.

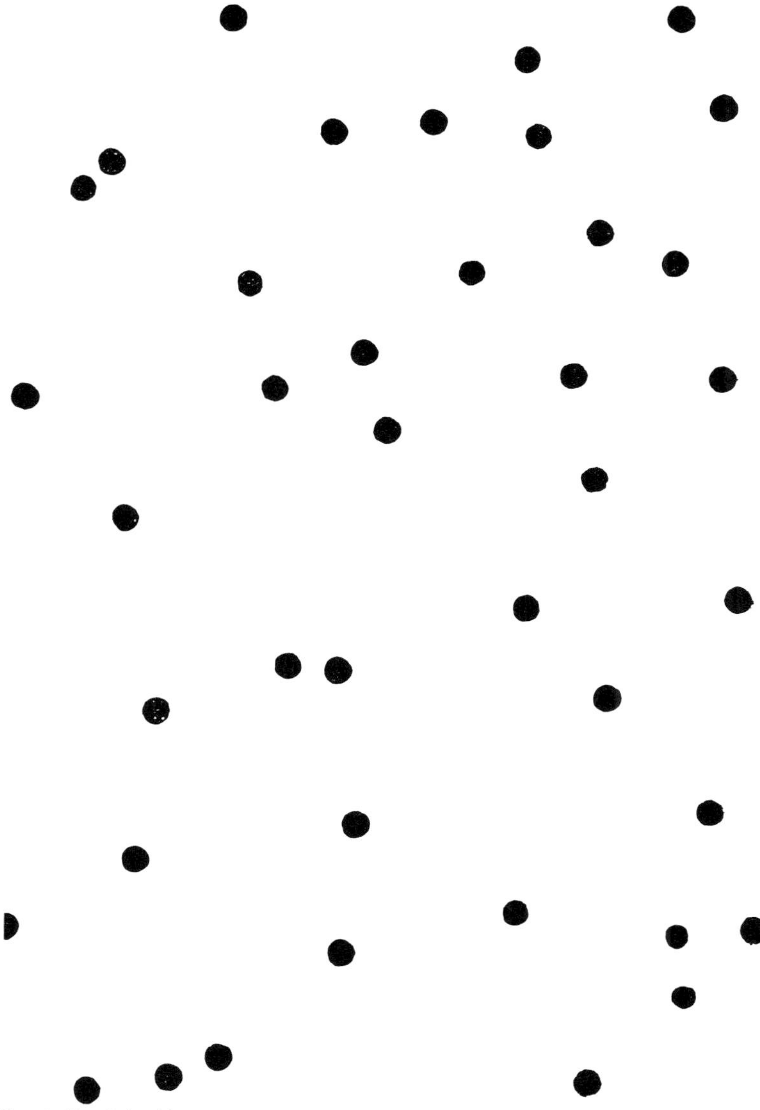

Fac-simile of the Shooting of a Full Choke-bored Gun at 40 Yards, with 3 drams and $1\frac{1}{8}$ oz. of No. 8 Shot.

LEFT BARREL.

Patterns made on 2nd July, 1885, with this Gun and W. W. GREENER's Loaded Cartridges, 42 grains of Schultze Powder and $1\frac{1}{8}$ oz. No. 6 Chilled Shot (305 pellets counted in).	273 276 255 276 260 252
Average	265·3

LEFT BARREL.

Patterns made with same Gun on 30th July, 1885. W. W. GREENER's specially Loaded Cartridges, 42 grains of Schultze Powder, and $1\frac{1}{8}$ oz. of No. 6 Chilled Shot (305 pellets counted in).	259 251 241 235 240 260
Average	247·4

Average obtained before two Gentlemen of the "Field" Staff, Aug. 8th, 1885—255.

This gun—as should all guns—shoots all first-class powders equally well. At a gun trial held at Leavenworth in 1886, a Greener 12-bore gun was shot with "King's Quick Shot" powder, an explosive the author has never had an opportunity of trying. The gun beat all its opponents easily—some were much heavier guns and of larger calibre—and made a target of which the appended diagram is a fac-simile.

These diagrams have been disputed more than once, but there is overwhelming proof as to their accuracy. They have been reproduced by a photographic process, the fidelity of which cannot be doubted.

In the specification of what a gun should be and do, the reader will find the patterns of various guns with various charges detailed, and it will only be necessary to add here that all patterns should be calculated upon the average of at least twelve shots, and that the frequent recurrence of a patchy pattern should condemn the gun. There must be absolute uniformity and regularity in pattern from a gun of the first quality. A very regular pattern is to be chosen, even before a *very* close one—if the closer pattern is patchy or not always good alike.

The two preceding diagrams will give the reader an idea of the actual closeness of pattern in the centre of the target. The one with No. 6 shot is equal to a pattern of 230 in a 30-inch circle; the one with No. 8 is equal to a pattern of 300 in the 30-inch circle.

KILLING CIRCLES.

The killing circle is the spread of the charge from the centre of the line of flight. It is apparent that the larger the killing circle and the greater the range the greater the efficiency of the gun.

At forty yards from the muzzle of a gun it has been proved that on frequent occasions a few pellets of the charge will be found 10, 15, and even 20 yards from the centre of the body of the charge; thus, at 40 yards a gun may, whilst putting the greater number of its pellets into a 30-inch circle, scatter some forty yards asunder.

The following fac-simile reproduction of targets made by the author will enable the sportsman to see at a glance the comparative density of patterns, and the approximate killing spread of the gun. These targets, obtained with guns of different gauges, may be approximated by guns of any gauge by altering the load or the range, or both.

Number of pellets in circle, 163. Killing circle about 26 in. Diagram represents the shooting of a 28-bore gun full-choked, at 40 yards, with $1\frac{1}{2}$ drams powder, and $\frac{7}{8}$ oz. No. 7 shot.

A similar pattern would be made with a 20-bore, and 1 oz. No. 6 shot, or a 20-bore with $1\frac{1}{8}$ oz. No. 5 should be no closer, but a killing circle two inches larger.

1

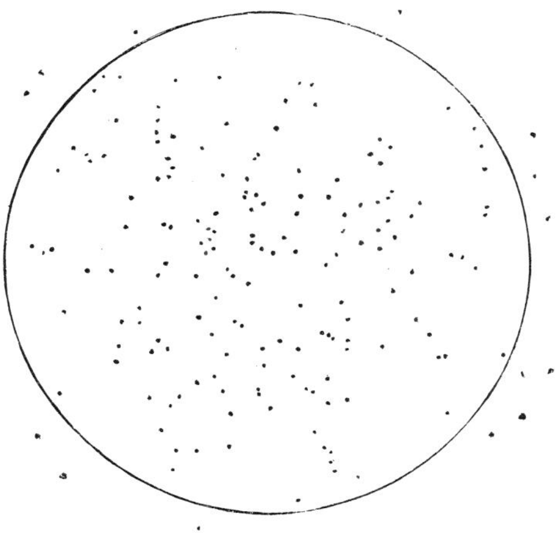

Fac-simile No. 1.—Circle, 30-in. diameter.

No. of pellets in circle, 285. Killing circle 30 in. This diagram represents the shooting of a 28-bore gun cylinder, at 20 yards with $1\frac{1}{2}$ drams, and $\frac{7}{8}$ oz. No. 7 shot.

A similar result is attainable from a 20-bore cylinder with $\frac{3}{4}$ oz. No. 8 shot.

2

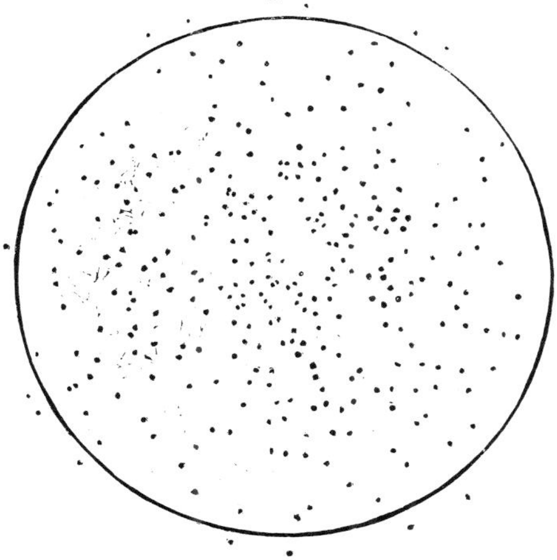

Fac-simile No. 2.—Circle, 30-in. diameter.

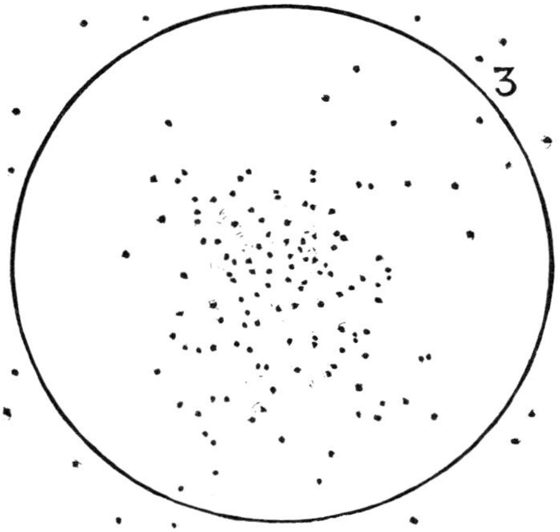

Number of pellets in circle, 131. Killing circle about 18 in. This diagram represents the shooting of a 28-bore gun, choke-bored, at 20 yards distance; charge, $1\frac{1}{2}$ drams, and $\frac{3}{4}$ oz. of No. 6.

A similar pattern results from using a 20-bore with 1 oz. No. 5 shot at 18 yards, with $\frac{7}{8}$ oz. No. 6 at 20-bore at 20 yards, makes a killing circle about two inches larger.

Fac-simile No. 3.—Circle, 30-in. diameter.

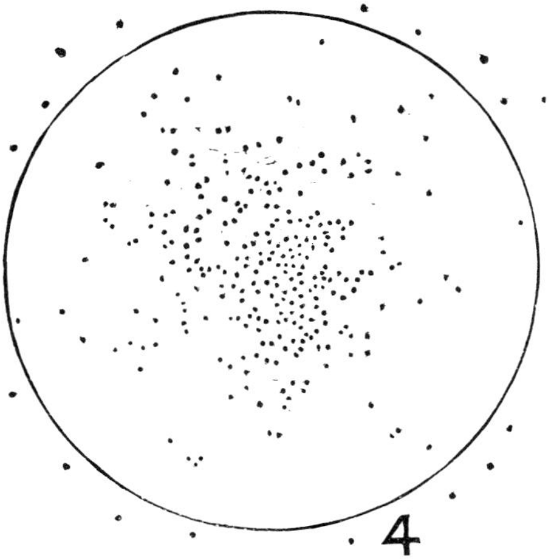

Number of pellets in circle, 292. Killing circle about 25 in. This diagram represents the shooting of a 12-bore gun choke-bored; distance 20 yards; charge, 3 drams and $1\frac{1}{8}$ oz. No. 6 shot.

A similar pattern results with a 20-bore at 20 yards with 1 oz. No. 8 shot, but with the 20-bore the killing circle is a little less.

Fac-simile No. 4.—Circle, 30-in. diameter.

No. of pellets in circle, 288. Killing circle, 30 in. This diagram represents the shooting of 12-bore gun cylinder at 20 yards; charge, 3 drams, and $1\frac{1}{8}$ oz. No. 6.

The same result is obtainable from a choke at 20 yards, by using 1 oz. No. 6 and *scatter charge*, or by using a brass case gun at 40 yards with $1\frac{1}{4}$ oz. No. 7, or with $1\frac{1}{8}$ oz. No. 8 at 40 yards.

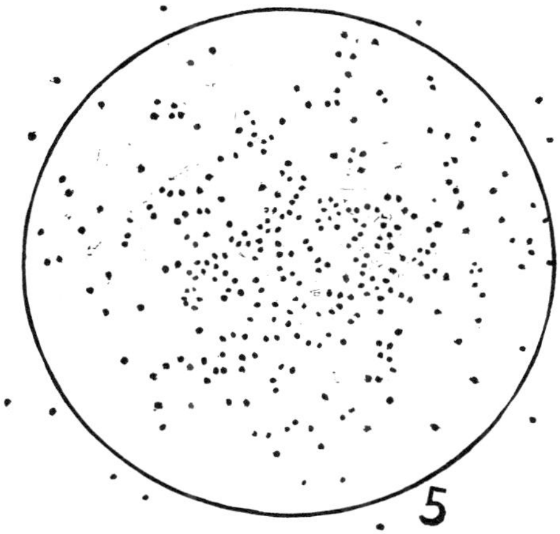

Fac-simile No. 5.—Circle, 30 in. diameter.

Number of pellets in circle, 250. Killing circle, 35. This diagram represents the shooting of a pigeon gun, 12-bore, with 4 drams and $1\frac{1}{4}$ oz. No. 6 shot.

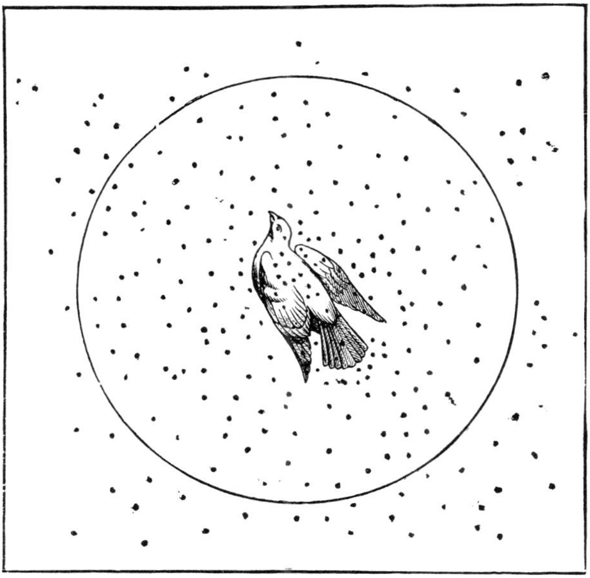

Fac-simile No. 6.—Circle, 30 in. diameter, Plate 4 foot.

The boring for a gun to shoot as No. 6 fac-simile is of a special kind, designed to produce a regular pattern not too thick in centre, but sufficiently thick to kill in a circle of 35 inches.

The instructions for approximating any other gun to one of the depicted patterns are based upon several series of experiments made at different times, and the data are sufficient to permit of reliable conclusions.

The *best pattern* is that of the choke-bore gun, and sportsmen seem slow to grasp the fact that pattern is the all-important factor in the killing range of the gun. The author made a series of experiments by which he ascertained that it requires at least four pellets of No. 6 shot (chilled) to kill a pigeon, excepting, of course, such flukes as a pellet striking the head or breaking the neck, and the pigeon, if struck by six shots, none of which might prove immediately mortal, will be dropped at once by the aggregate. A cylinder gun will not, upon the average, put three pellets into a pigeon 30 yards distant. The cylinder gun must, therefore, be considered practically useless at this distance; for, providing the pigeons were fairly struck, and in the centre of the charge, not more than one out of three would be killed and gathered. To ensure four pellets being put into a pigeon, a pattern of at least 200 in the 30-inch circle is necessary. It is possible to kill pigeons at 40 yards with a choke-bored gun if so small a charge as $2\frac{1}{4}$ drams and $1\frac{1}{4}$ oz. of No. 6 shot be used; with a pattern of 240 in the 30-inch circle the value of penetration is lessened. Whereas, unless the pellets *strike* the bird, penetration is quite useless; pattern never is.

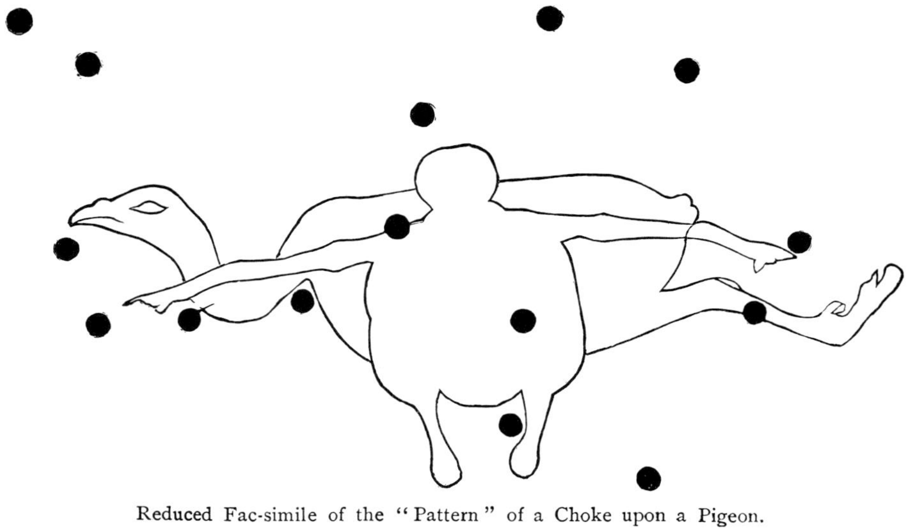

Reduced Fac-simile of the "Pattern" of a Choke upon a Pigeon.

The cylinder gun must not be used at distances greater than 30 yards; to do so is unnecessarily cruel, for the author has known as many as six successive shots to have been fired from a cylinder gun at a stationary pigeon without killing it, the distance being only 35 measured yards, the charge used a full one, and the pigeon well in the centre of the pellets each time. After the sixth shot the bird was examined, and it was found that it had been struck by only nine pellets.

A bird flying directly away from or towards the shooter would have been struck by a still smaller number of shot. A cylinder-bore gun at a range of 15 yards has put 54 pellets into a pigeon. A choke at 20 yards range will average only 40. The annexed rough outline of a pigeon in flight end on and across the charge will show by how many pellets the bird would have been struck. These outlines were sketched in the sheet before firing at it with a 12-bore choked gun, and the illustration is an exact reproduction of the target.

Continuing this theme still further, the small bore will be found to kill quite as well as the larger bore, providing the pattern be as close, but when the bird struck is not in the centre of the charge, it is not often killed.

A pigeon placed in a wooden box measuring six inches by seven inches, just sufficient room being allowed for it to stand, with its broadside to the gun, a piece of thin paper only between the bird and the gun, was fired at various ranges with a 28-bore gun with the following results:—

				CHARGE.			PATTERN.	RESULTS.
				Dram.			ON 7 × 6	
No. 1	Pigeon,	distance	40 yards	$1\frac{1}{2}$	$\frac{3}{4}$	No. 6. ...	13	Bird struck in body, but not in any way disabled.
No. 2	,,	,,	40 ,,	$1\frac{3}{4}$	$\frac{3}{4}$	,, ...	12	Leg broken, and 1 pellet in breast, not disabled from flying.
No. 3	,,	,,	35 ,,	1	$\frac{3}{4}$	,, ...	18	Hit in body, but not disabled.
No. 4	,,	,,	35 ,,	$1\frac{1}{2}$	$\frac{3}{4}$	,, ...	18	Do. Do.
No. 5	,,	,,	30 ,,	$1\frac{1}{2}$	$\frac{3}{4}$	,, ...	23	Killed dead.
No. 6	,,	,,	30 ,,	1	$\frac{3}{4}$	,, ...	35	Killed dead.

IGNITION.

Different makes of cartridge cases, being furnished with different caps, are not alike in the method of ignition, of the explosion, nor in the time occupied in igniting the powder.

The "Life" cases have the largest and best caps, and ignite all powders more quickly than the usual English cap.

The stronger ignition of the "Life" case led the author to suppose that it might be attributed to some of those dangerous explosions in small-bore guns with Schultze powder, that have so astonished gunmakers and sportsmen the past two seasons. The editor of the *Field* kindly experimented to determine this question, and declared that although there was a difference in the strength exhibited by Schultze, when fired by the larger cap, the increase was so infinitesimal that it would not account for the disastrous results partly attributed to its use.

The time actually passed between the pulling of the trigger of a good C. F. 12-bore gun, and the exit of the shot from the muzzle of the gun, should be ·0075 of a second; with the "Life" cases this is reduced to ·0070 of a second, and if quick powder, as "basket" or No. 3, is used, it is again reduced to ·0064 of a second. With smaller bores the time is less. When the time taken is ·03 of a second or more, a "hang fire" is perceptible; when ·06 of a second a click is heard between the pulling of the trigger and the report.

The questions of ignition and combustion of various explosives are treated of more fully in the second part of this treatise.

SECTION V.—SPECIFICAL.

THE ESSENTIALS OF A GUN.

THE purposes for which a sporting gun is required are various. The author has been commanded to build one gun which shall be effectual at all game, from snipe to an elephant; and although this weapon was tolerably successful at everything, it is not to be supposed that it was actually fitted for anything except the elephant. Another customer of the author's, to decide a bet, shot a couple of snipe with a heavy elephant rifle of 8-bore. There is, therefore, no actual limit to the capabilities of any weapon until trial has been made. The collector who requires humming-birds, and the wild-fowler who thinks of getting wild geese, will arm themselves very differently.

Again, some guns have to be carried throughout a long day's walk; in other sports the gun is only in the hand the couple of seconds requisite to aim and fire. It is, therefore, evident that what is desired for one sport is of little importance in a gun desired for another sport.

There are certain essentials, however, which should be possessed by all varieties of guns. Amongst the chief of them are:—Facility in loading at the breech, freedom from danger to the user or his companions, simplicity of mechanism, speed in manipulation, handiness, lasting power.

The ordinary game gun should have a killing circle of 30 inches at thirty yards with the first barrel, and at forty yards with the second.

The gun for covert shooting will give a 30-inch killing circle at twenty yards with the first and at 30 with the second.

The gun for grouse driving will be bored to give a killing circle of thirty inches in diameter at the longest possible range; the gun not to be more than $7\frac{1}{2}$lbs. weight; and it is not fashionable to use a larger bore than 12 for this sport.

The gun for pigeon shooting must be so built as to meet the rules of the chief clubs; in England the bore must not be larger than 12, nor the gun heavier than 8lbs.; the charge to be used must not exceed four drams of powder and $1\frac{1}{4}$ ounces of shot. On the Continent and in America 10-bores are allowed, but there is usually some restriction as to charge. The pigeon gun may be made with hammers or hammerless, preferably the latter. It should *not* have a trigger bolting safety, and an automatic trigger safety for this species of gun is the greatest mistake that can be made.

The shooting required will in some measure depend upon the distance at which the user is generally placed, it being required to have the largest possible killing circle at one yard beyond the trap with the first barrel and at five yards with the second. In no class of gun is uniformity and regularity of shooting more essential than in the trap-gun. The weight may be from $\frac{1}{2}$ to $\frac{3}{4}$ of a pound greater than in the gun carried for game shooting, but it is important that the balance be perfect.

Light guns of small calibre, such as generally placed in the hands of boys, can be made formidable game weapons. Even the 28-bore, generally considered a toy, is in the hands of a good shot a serviceable sporting gun. "Young Nimrod," when 11 years of age, shot with a 28-bore of the author's make, and did remarkably well. In public pigeon matches he was placed at 27 yards, and at that distance upon more than one occasion has killed

The Pigeon Gun.

his 38 out of 50 best Blue Rocks. Sometimes he would grass many birds in succession, several strayings of 17, 11, and 13 having been scored to him, which is evidence not only of his skill as a marksman, but of the killing powers of his gun.

A nobleman, well known in sporting circles, wrote the author the following in November, 1884 :—

"I had the 28-bore out for a few shots at pheasants yesterday, and I am much pleased with it, killing eight birds in succession, and four of them at least thirty-five yards off, flying away low, and one with the choke-barrel a very long shot—we measured it—fifty-three yards, and the bird was flying away within three yards from the ground; it fell stone dead to the gun. I shot a hare with the right—not choked—barrel at thirty-four yards as dead as a nail. (Charge used, $1\frac{1}{2}$ drams black powder $\frac{3}{4}$ oz. No. 6)."

Again, on February 4, 1885 :—

"I can only say your 28-bore gun cannot be improved upon; its shooting is quite first-class. I have given it a capital trial, and find it shoots as strong as a 12-bore. Of course you have to lay on straight, *then* I defy any gun to shoot harder; it has had a really good and heavy trial."

Another good shot writes :—

"I have tried the little gun, have made some very long dead shots at rabbits, and am confident of success at game."

It is therefore quite indisputable that these small-bore guns are really efficient weapons in the hands of good shots; their lightness recommends them; and although they seem toy-like, they may afford as good sport, and fill the bag as well as larger and more cumbersome weapons.

A single gun made for a lady, and weighing but a trifle over three pounds, is a first-class all-round game gun; partridges, pheasants, hares, and rabbits are shot regularly. The efficiency of the gun, however, is better demonstrated by the fact that it is preferred by the owner's brothers to their own 12-bore guns for shooting at the wood pigeons as they come home to roost in the high elms in the park, and on one occasion a fallow doe was shot dead with it at 25 yards' distance with seven-eighths of an ounce of No. 7 shot.

The small bores may therefore be ranked as serviceable weapons, whilst for boys about to commence shooting, the 28-or 24-bore double is to be preferred to the single gun. They are, of course, more expensive; to build them well requires more care and a greater outlay than the building of a gun of ordinary sporting gauge.

The Ladies' Gun should be purposely constructed; not only must the stock be differently shaped and of very different measures to the ordinary gun, but the barrels will require modification if the best possible results are to be obtained.

There are some sportswomen who can shoot well with almost any gun, just as there are men who use guns of divers bends and weights indifferently, but to most ladies the question of recoil is an important one. The author, having had more experience in the building of guns for ladies' use than perhaps any other English gunmaker, can confidently assert that the gun possessing the essentials he is about to enumerate will prove more effectual than the light small-bore guns usually recommended.

The bore 12, the barrels twenty-seven inches long, the weight $5\frac{3}{4}$lbs., making with the right barrel a killing circle of 30 inches at 30 yards, with the left a similar pattern at 35 yards, the charges to be used being in the right, $2\frac{3}{4}$ drams of black or 36 grains of "Schultze" or "E.C." powder and 1 oz. of No. 7 shot, and in the left barely 3 drams by measure of black or 40 grains of "Schultze" or "E.C.," and $1\frac{1}{8}$ ozs. of No. 6 shot. The stock to be suitably shaped, well bent, and well cast off; the gun to be perfectly balanced; and not butt-heavy. A 12-bore gun cannot be made satisfactorily to weigh less than $5\frac{3}{4}$ lbs.; if a lighter gun be required a more serviceable weapon will be obtained by choosing a smaller bore, viz., 16, or even 20.

This is the Miniature Gun, a weapon which has come rapidly into favour, one of its first and strongest advocates being that well-known authority upon shooting matters, "One who has fired 20,000 shots at marks." The principle is that of reducing the 12-bore gun in length and bore of barrel to the exact capacity required by the sporting charge of 3 drams and $1\frac{1}{8}$ ounces. From numerous experiments the author has arrived at the conclusion that a barrel of 25 inches long, choke-bored, and of a certain gauge, twenty inches of its length will satisfactorily burn 3 drams of powder, and propel $1\frac{1}{8}$ ounces of shot at a high velocity: in fact, that the 12-bore gun is "too much gun" for the charge of 3 drams and $1\frac{1}{8}$ ounces, that is to say, there is too much gun in proportion to the charge of powder than absolutely necessary; for ordinary game shooting the 12-bore with 25 inches will shoot this charge as well as it need be shot. By carefully reducing the 12-bore gun, however, in barrels, breech-action, locks, and stock, a miniature gun is produced from one to one-and-a-half pounds lighter than the normal 12-bore, and shooting the standard 12-bore charge nearly as well as the ordinary 12-bore choke gun does. These miniature guns require great care, and very considerable tact to be exercised in their manufacture, and it is quite impossible for any maker without practical experience to produce perfect weapons of this kind. The 27-inch barrels will be found to permit of better marksmanship than shorter barrels, and, consequently, unless there is a good reason for doing so, guns should not be made with barrels shorter than 27 inches. Although they are sometimes made lighter than $5\frac{1}{2}$ lbs., it is only at a sacrifice of strength. A reliable gun with breech ends of the barrels of the ordinary thickness can be made as light as $5\frac{3}{4}$ lbs., below which it is inadvisable to go.

A miniature 12-bore gun, therefore, will always command a fair price, and can never be made in the cheapest grades. It must fire 3 drams and $1\frac{1}{8}$ ounces to perfection, and without appreciable recoil—a larger charge cannot be used with comfort—balance and handle perfectly; *every part* being reduced from the ordinary 12-bore gun size, it must stand the heavy wear and tear of the hardest season, and yet be perfectly safe.

This is the weapon Birmingham has produced, and its many advantages will commend it to those sportsmen whose work is not such as lies beyond the capabilities of three drams of powder and $1\frac{1}{8}$ ounces of shot.

The Miniature Gun is much to be preferred to the Single Gun, a species of shot gun quickly falling into disuse. The double is now constructed so light that a Single Gun, if made lighter, would recoil unbearably. It is for duck guns and large-bore rifles that the single barrel is mostly used. Two heavy barrels of 4-bore, side by side, are more than the hand can firmly grasp, so many shooters adhere to single guns for wild-fowling, preferring to lose the chances of a second shot than possess only an imperfect command of the gun.

The Duck Gun may be a double 10-bore of 10 or 11lbs. weight, firing 5 drams of powder and 2 ounces of shot with a killing range of 90 yards with large shot; but the true wild-fowling gun must be 8 or 4-bore, with a long barrel, and firing a heavy charge, and specially regulated for large shot.

A double 8-bore with 34-inch barrels should weigh 14lbs., and is an excellent weapon for shore shooting.

A light 8-bore gun is much to be preferred as a "duck" gun to any 10-bore gun. The Duck Gun, for use from the shoulder, recommended by

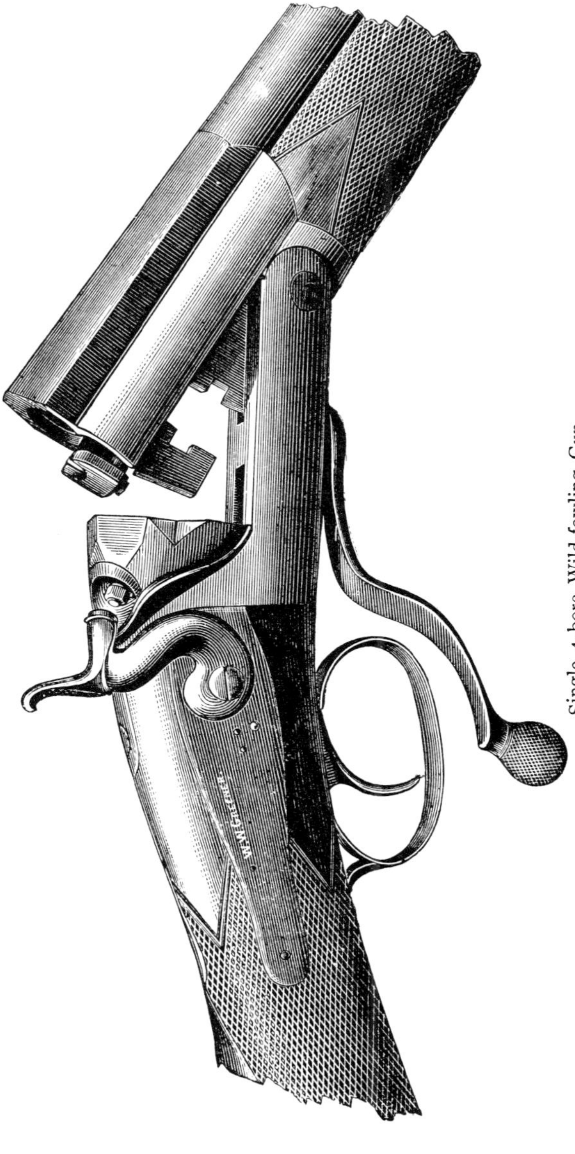

Single 4-bore Wild-fowling Gun.

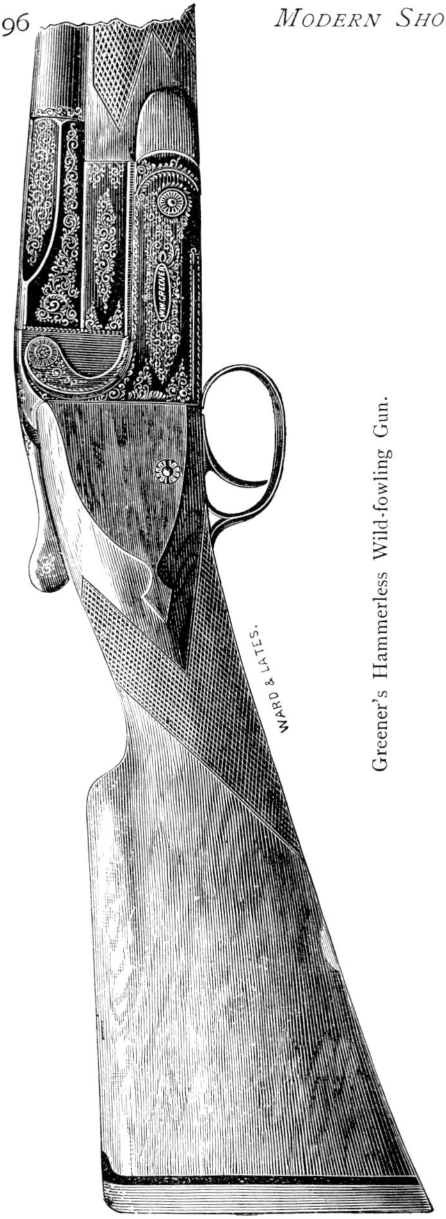

Greener's Hammerless Wild-fowling Gun.

the author, is an 8-bore, with 32-inch barrels, and weighing from $10\frac{1}{2}$ to 11 lbs. Such a gun will shoot 6 drams of powder and 2 ozs. of shot to good advantage, and if used with large shot will show such a decided superiority over 10-bores that sportsmen cannot fail to appreciate the difference. This gun, with No. 4 or No. 2 shot, will shoot more closely than a 10-bore gun, and enable the sportsman to bag more birds. Its weight is not more than that of some 10-bores used by practised wild-fowlers, the recoil is not heavy, and the pattern and range are superior to anything obtainable from any guns of the same weight.

The single 8-bore with 36-inch barrel may weigh 15 lbs.; the single 4-bore may have a barrel of 42 inches, and weigh 18lbs. These are true wild-fowling guns.

The cheapest breech-loading gun for wild-fowling is made on the double-grip principle, as illustrated here; it is a serviceable weapon, and will cost from £15 to £30 in 8-bore, and from £20 to £35 if 4-bore.

The newest style is hammerless, with top-lever cross top-bolt. This gun has greater speed, is safer and stronger, is neater in appearance, and in many ways is better than the double-grip hammer gun. It is more expensive, costing, when well made, not less than £30, and in best quality, £45.

All wild-fowling guns should be specially bored and chambered, for solid brass, drawn or "perfect" brass cases, which possess many advantages over paper for wild-fowling. All wild-fowling cartridges are heavy and cumbrous, but in the brass there is more room for a larger charge, in the same length as the paper; the escape of gas at the base is an impossibility with brass cases; therefore they often shoot more strongly, and they will not jamb in the chamber. Changes of atmosphere do not affect them to the extent paper cases are affected.

It is needless to say that all duck and wild-fowling guns should be choke-bored to the fullest extent. In choosing a wild-fowling gun, it must be borne in mind that 8-bores shoot much better in proportion than do 10-bores to 12-bores, whilst 4-bores are only as much better in proportion to 8-bores as 10-bores are to 12-bores.

With a first-class 12-bore it is possible to get patterns of 255 in a 30-inch circle at 40 yards, and with the same charge; and a 10-bore gun it is not often a better pattern is made, and 275 is certainly an excellent average for a 10-bore duck gun, with the 10-bore standard charge. When shooting large shot, the 10-bores show a more marked superiority over the 12-bores.

The 10-bore duck-gun full-choked, weighing $8\frac{1}{2}$ lbs. and over, will make a pattern of 275 in a thirty-inch circle at forty yards, with 4 drams of powder and $1\frac{3}{8}$ ozs. of No. 6 shot. This pattern is as represented in the accompanying fac-simile diagram of a portion of the circle.

With $4\frac{1}{4}$ drams, and $1\frac{1}{2}$ ozs. of No. 2 shot, pattern in 30-inch at 40 yards, 160 pellets, penetration 25 sheets of strawboard. Same charge, distance, &c., but No. 1 shot, pattern 135, 50 being in a centre of 12 square inches; all pellets should be in a 25-inch circle at this distance—penetration 31 sheets of strawboard. At 60 yards, No. 4 shot, the pattern should average 78, and penetration be 18 sheets, with No. 1 shot, the pattern 60, the penetration 26 sheets.

The 8-bore with brass cases, standard charge 7 drams and $2\frac{1}{2}$ ounces of shot. The light 8-bore with 32-inch barrels is suitable for a charge of 2 ounces of No. 6 shot; and with No. 6 shot should put about 400 pellets in a 30-inch circle at 40 yards. The 8-bore is, however, at its best with shot of numbers between and including 4 and 1. With No. 1 a sufficient velocity is attained to kill ducks at 150 yards, and a certainty of killing at 100 yards if the gun be held straight; at greater distances the pattern is necessarily thin, and there will always be a risk of the bird escaping.

With $2\frac{1}{2}$ ozs. No. 4 shot at 40 yards, the pattern should average 300, and the penetration be equal to 34 sheets of strawboard; with No. 1 shot, pattern 220, penetration 40 sheets; at 60 yards the penetration of No. 1 shot 34 sheets, at 80 yards 24 sheets, at 100 yards 16 sheets. The penetration of seven sheets by one shot is sufficient to kill a duck.

With paper cases, 3 drams and $2\frac{1}{2}$ ozs. of No. 1 shot, has given a pattern of 195 pellets, at 40 yards; same charge and conditions, but brass cases, the pattern was 225.

The 4-bore brass case gun is quite as powerful as the 2-bore paper case

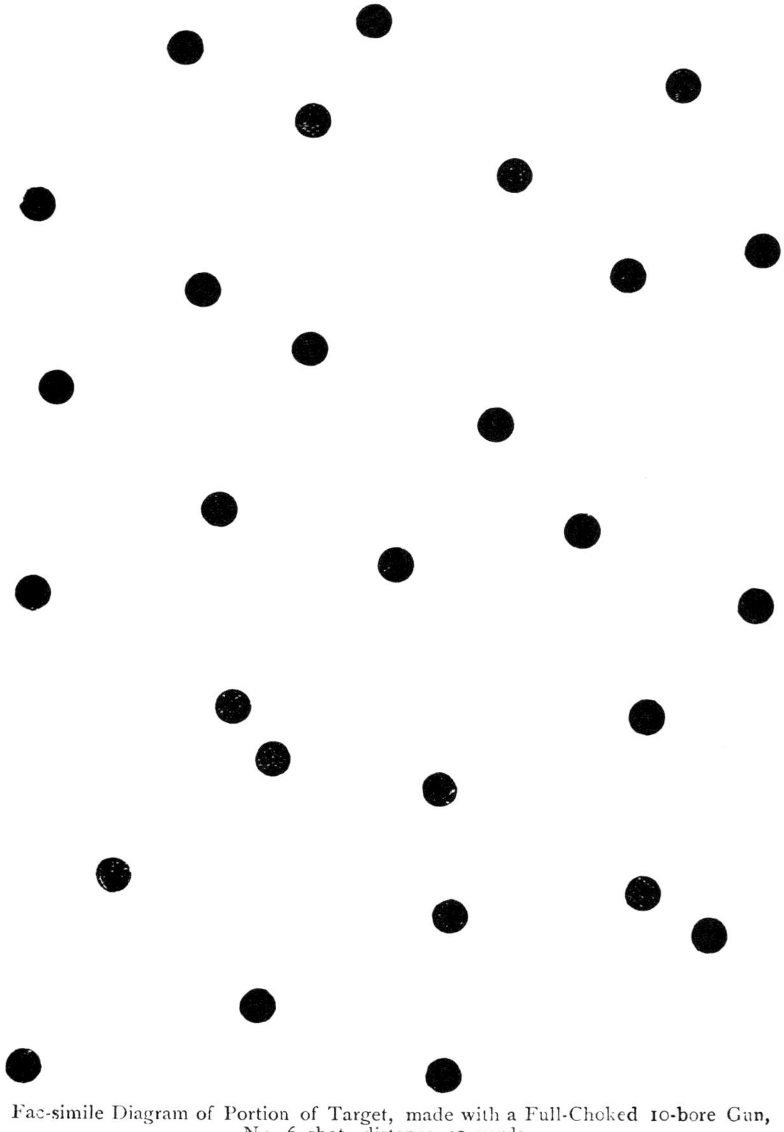

Fac-simile Diagram of Portion of Target, made with a Full-Choked 10-bore Gun, No. 6 shot, distance 40 yards.

Fac-simile Diagram made by Greener's 4-bore Wild-Fowling Gun, with No. 1 shot.

gun—the largest sporting bore made. A fac-simile diagram of the shooting of a good 4-bore is appended.

The standard charge is 9 drams and 3½ ozs. This charge and No. 1 shot will average a pattern of 243, and penetration of 46 sheets at 40 yards; at 60 yards, pattern 150, penetration 34 sheets; at eighty yards, pattern 65, penetration 24 sheets; at 100 yards, pattern 23, penetration 16 sheets. With 12 drams and 2 ozs. of B. shot, a pattern of 230, and penetration of 54 sheets at 40 yards. The 4-bore shoots large shots well to the centre, and gives a killing circle of about 6 feet at 100 yards.

The subject of wild-fowl guns, their many modifications, and the peculiar requirements of shore shooters and wild-fowlers generally, requires more space than the author can now command. The wild-fowler, however, will not hesitate to procure such books as "The Gun and its Development," the volume "Moor and Marsh," of the Badminton Library, and "Wild-fowling in Ireland," by Sir Ralph Payne Gallwey, in all of which modern works he will find his speciality more fully treated.

The brass case, as already stated, is admirably adapted for use in wild-fowling guns, but since the introduction of cheap brass cases, a special class of gun has been produced to use them to best advantage. These guns are generally known as Brass Case Guns.

The brass case gun is a gun in which thin brass cases alone can be used if good shooting is required. The difference between the diameter of the chamber and the barrel is so infinitesimal that the cone is practically abolished. Brass case guns have, therefore, larger barrels than paper case guns of the same nominal gauge. A 12-bore paper case gun is about 13-bore, the 12-bore brass case gun is nearly 11-bore. A 12-bore brass case gun, if built strongly and with plenty of metal in the barrels, may be used with 10-bore charges. The 14-bore brass case gun is equal to an ordinary 12-bore paper case gun. The various advantages of the two guns will be specified in a succeeding section.

WHAT A GUN SHOULD DO.

THE 28-BORE should have 25 or 27-inch barrels, which will require but little choking, and average :—

Charge powder.		30-inch Pattern.	Cardboard penetration.	Mean velocity.	Force at impact.
1½ drams	¾ oz. No. 8	150	13	705	0·66
*32 grs. Schultze	¾ oz. No. 6	124	22	940	1·96
1½ drams	¾ oz. No. 6	130	17	720	1·87

* Over-loaded. This charge has been too frequently used; 26 grains does better.

The weight should not be less than 4, nor more than 4¾ lbs. Recoil 60 lbs. The 28-bore must not be loaded with 1 oz. of shot, as is too often done. This calibre especially is too frequently much over-loaded. The 28-bore must be used with brass cases if the full capability of this calibre is desired.

THE 24-BORE is but little used; it comes about midway in pattern, penetration, recoil, &c., between the 28 and 20-bore.

THE 20-BORE is the smallest bore sought after by the general sportsman; a gun of $5\frac{1}{2}$ lbs. weight, and with 28-inch barrels, may be taken as representing fairly the 20-cal. class, and should average—

AT FORTY YARDS.

Charge.		Pattern.	Penetration of	Mean	Force at
Drams of powder	Oz. of shot.	30-inch circle.	Strawboards.	Velocity.	Impact.
$2\frac{1}{4}$	1 No. 8	230	13	708	0·67
$2\frac{1}{4}$	1 No. 6	180	18	725	1·37
$2\frac{1}{2}$	$\frac{7}{8}$ No. 6	160	19	734	1·94
$2\frac{1}{2}$	$\frac{7}{8}$ No. 5	120	22	742	2·37
$2\frac{1}{2}$	$\frac{7}{8}$ No. 1	65	29	815	2·72
32 grs. Schultze	$\frac{3}{4}$ No. 6	150	20	913	2·00
AT SIXTY YARDS.					
$2\frac{1}{4}$	1 No. 6	60	9	725	0·76
$2\frac{1}{4}$	1 No. 5	35	18	737	0·89

Weight not less than $5\frac{1}{4}$ nor more than 6 lbs. Recoil under 90 lbs.

THE 16-BORE FULL-CHOKE, with barrels 30 inches in length, and the gun weighing $6\frac{1}{4}$ lbs., should average—

AT FORTY YARDS.

Charge.		Pattern.	Penetration of	Mean	Force at
Drams of powder	Oz. of shot.	30-inch circle.	Strawboards.	Velocity.	Impact.
2	1 No. 6	180	17	713	0·83
2	1 No. 5	160	22	731	0·98
$2\frac{1}{2}$	1 No. 6	190	19	733	1·35
$2\frac{1}{2}$	1 No 5	155	23	745	2·49
$2\frac{3}{4}$	1 No. 6	180	21	741	2·27
3	1 No. 6	174	22	744	2·33
$2\frac{3}{4}$	1 No. 5	155	25	753	2·48
$2\frac{1}{2}$	1 No. 1	85	29	813	3·00
AT SIXTY YARDS.					
$2\frac{3}{4}$	1 No. 6	95	9	741	1·10
$2\frac{3}{4}$	1 No. 5	85	12	753	1·26
$2\frac{1}{2}$	1 No 1	45	19	813	1·52

Weight from $5\frac{3}{4}$ to $6\frac{1}{2}$ lbs. Recoil 82 lbs.

A 12-BORE FULL-CHOKE may be made as light as 6 lbs., with 27-in. or 28-in. barrels, but from 6 lbs. 10 oz. to $7\frac{1}{4}$ lbs. is the more general weight. Guns under 7 lbs. should have 28-in. barrels; over that they may be 30 in. The author has made 12-bores with barrels but 24 in. long, and they have performed remarkably well with ordinary charges; also 12-bores, with longer barrels—32, 34, and 36 in.—but their performance has not surpassed that of 28 or 30-in. barrels with ordinary charges.

Twelve-bores under 7 lbs. will not shoot a heavier charge than $3\frac{1}{4}$ drams and $1\frac{1}{8}$ oz. with comfort to the shooter. If over 7 lbs. and under $7\frac{1}{2}$ lbs., the charge may be $3\frac{1}{2}$ drams and $1\frac{1}{4}$ oz.; over $7\frac{1}{2}$ lbs., guns are usually built for extra-long cartridge-cases and special charges.

The usual full-choked 12-cal. gun with 30-inch barrels, and weighing 7 lbs., should average—

AT FORTY YARDS.

Charge.		Pattern.		Penetration of Strawboards.	Mean Velocity.	Force at Impact.
Drams of powder.	Oz. of shot.	Square 10″ centre.	30 in. circle.			
$3\frac{1}{4}$	$1\frac{1}{8}$ No. 8	92	320	15	800	0·80
3	$1\frac{1}{8}$ No. 6	55	215	20	839	1·80
$3\frac{1}{4}$	$1\frac{1}{8}$ No. 6	51	210	21	857	1·83
$3\frac{1}{2}$	$1\frac{1}{8}$ No. 6	39	200	22	912	1·87
$3\frac{1}{4}$	$1\frac{1}{4}$ No. 6	58	240	21	864	1·85
3	$1\frac{1}{8}$ No. 5	35	175	22	878	1·89
$3\frac{1}{4}$	$1\frac{1}{4}$ No. 5	45	190	23	882	1·91
$3\frac{1}{4}$	$1\frac{1}{4}$ No. 4	40	160	24	900	2·81
$3\frac{1}{4}$	$1\frac{1}{4}$ No. 3	38	135	25	950	3·16
$3\frac{1}{4}$	$1\frac{1}{4}$ No. 1	35	105	26	980	4·18
$3\frac{1}{2}$	$1\frac{1}{4}$ No. 1	33	100	30	988	5·30
			AT SIXTY YARDS.			
3	$1\frac{1}{8}$ No. 6	29	110	10	839	0·93
$3\frac{1}{2}$	$1\frac{1}{8}$ No. 6	36	100	11	912	1·28
$3\frac{1}{2}$	$1\frac{1}{8}$ No. 5	28	89	14	926	1·47
$3\frac{1}{2}$	$1\frac{1}{8}$ No. 4	17	70	18	962	2·00
$3\frac{1}{4}$	$1\frac{1}{4}$ No. 1	15	50	21	980	2·46

LONG RANGE GUNS.

THE 12-BORE GUN is not made a more powerful weapon by lengthening the chambers, and using cartridge cases more than $2\frac{3}{4}$ inches in length. Even with the $2\frac{5}{8}$-in. case, a charge of 50 grains of Schultze or "E.C." powder may be pressed in, and with the full complement of wads, and $1\frac{1}{4}$ oz. of shot, will be found to give as good results as any obtainable with larger charges from this gauge.

Cases 3 inches in length have nothing to recommend them, and guns constructed for them require great attention, or they will not shoot at all regularly.

To make the 12-bore gun a more powerful shooter than its calibre warrants has long exercised the ingenuity of inventors. The Lancaster concentrator for increasing the range of guns was one of the earliest and perhaps one of the best of devices for the purpose, and has enabled long shots at wild fowl, with small and other bore guns, to prove efficacious where otherwise they would not have been so. The wire cartridge tended to still further concentrate the shot, and at ordinary ranges the charge too often fled as a bullet.

John's Shrapnell shell is one of the latest attempts to make a cylinder-bored shot gun efficient at very long ranges. The adjoining illustration shows this shell in both its forms.

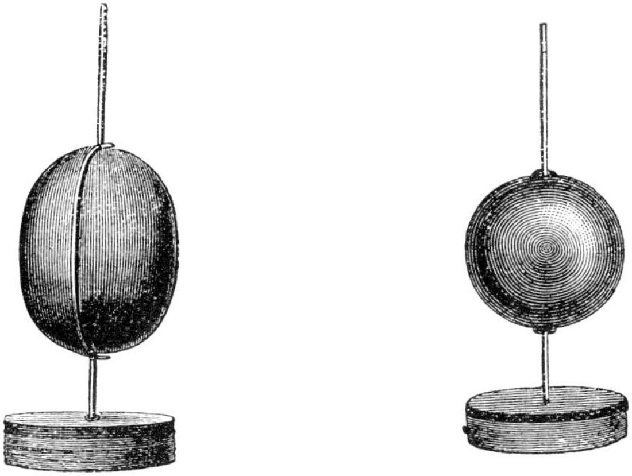

The Shrapnell Spherical and Elongated Shot Shells.

The Shrapnell shell consists of two segments, bolted together by a light wire spindle, its lower end being fixed to ordinary wadding slightly larger in circumference than the shell. The shell is put into the cartridge case instead of loose shot.

This shell differs from ordinary concentrators, and, according to the inventor, the following is the theory of its action :—The opening of the shell is regulated by the length of the spindle, yet for a certain period during its flight the spindle is locked in the shell owing to the temporary distension of the two segments by the contained shot pressing to the circumference of the shell. When the contained pellets acquire a uniform velocity, they settle down, the shell relaxes, and the spindle is snatched away. When the atmosphere begins to withdraw the spindle it is then that the length of spindle tells, and by the time the spindle is withdrawn from the shell the pellets have all acquired their own path, and cease to jostle each other ; at 120 yards a circle four feet in diameter should include the whole of the pattern. Owing to the temporary locking of the spindle the spherical shell has a minimum range of 90 yards : that is to say, it flies as a bullet for that distance, but it is effective at from 95 to 140 yards.

The Shrapnell shells and other concentrators are only of use in guns cylinder bored.

GUNS FOR BUCKSHOT.

A special gun is required to shoot large shot of three, four, or five to the layer, with the best results, and such guns of 12-bore, if correctly constructed, will shoot at long ranges with such force and accuracy that they may with advantage be substituted for rifles for small deer shooting.

The following letter, which appeared in the *Field* on February 15th, 1887, will convey an accurate idea of the nature and power of a true buckshot gun :—

SIR,—About a year ago I sent you an article, in which I gave my idea of boring a breech-loader for shooting buckshot with a certainty of killing a deer from 100 to 150 yards ; also stating if I could get W. W. Greener to make an extra pair of barrels for the gun he made for me several years since, and bore it according to my suggestion, I believed I could accomplish my object. In a short time after the article appeared in the *Field*, Mr. Greener wrote to me "that he fully understood what I wanted, but that it would be cheaper for me to have him build a new gun." I consented, and he sent me a No. 12 hammerless gun—30 in. barrels, weight $7\frac{1}{2}$ lbs.—which I received last June. It has more than met my most sanguine expectations, and fully verified my opinion, not only shooting buckshot with the certainty of killing a deer from 100 to 150 yards, but also proving a remarkable shooter with small shot.

During the past summer I only got shots at six deer, killing each shot. The longest shot was 91 yards, the deer being struck with three shot, one breaking the back, and the other two just below. I shot too high, the deer being in the act of leaping high, and the remaining six shots passed into a gum-tree above the height of the deer.

In a number of trials at a 30-in. circle, from 100 to 156 yards, not a shot was fired that would not have killed a deer. At the distance of 156 yards a shot made by Dr. Hargrove, of Knox Point Post-office, three shots would have entered the side of a deer.

Near Houghton Post-office a number of shots were fired by John A. Skannall, Money Brian, and George Conway. These gentlemen are among our most eminent planters, and distinguished for their fine shooting. At 100 yards from five to six shots were put in the target out of a possible nine at every discharge. At 125 yards never less than four shots would have entered the side of a deer.

Last week, while hunting partridges, I inserted a shell loaded with buckshot, and gave it to Mr. Tom Barrett, a lawyer of Shreveport, to shoot at a sparrow-hawk. He killed it at the distance of 90 yards.

At a trial made in Shreveport a number of distinguished gentlemen attended. They were sceptical as to the long range of this gun, and would not believe, unless they measured the distance, and shot the gun. They brought a tape line and measured 125 yards. Among them were Hon. A. C. Blanchard, member of Congress from the Fourth District of Louisiana ; his law partner, Alexander ; and Capt. Smith, superintendent of the Fair Grounds. They themselves shot my gun with results similar to what I have above stated, putting from two to three shots in less than 3 in. of the centre of a 30-in. circle at every discharge. I give the names of these gentlemen, with post-office addresses, so that, should any one doubt my statement, they could be referred to. I would also add the names of Capt. Ike Dyer and Capt. Jas. Y. Webb, of Minden, La., who were the first to test the qualities of this gun after I received it.

It is a very great advantage to have a breech-loader doing such extraordinary shooting with buckshot, and at the same time proving a very fine gun with small shot.

I have no doubt that W. W. Greener, of St. Mary's Square, Birmingham, England, could duplicate this gun for any one who may desire to get the best deer gun which has been manufactured during this century.

Not long since, in the presence of a number of cadets of the Thatcher Military Institute, of Shreveport, I fired with buckshot at a 3-in. circle, 90 yards distant, and struck it with three shots, one grazing the centre. A deer would have been struck with eight of the possible nine shots.

GEO. D. ALEXANDER.

SHOT-GUNS AS BALL-GUNS.

It is well known that the ordinary double-barrelled cylinder shot-gun will shoot spherical bullets with fair accuracy up to fifty yards.

The recoil felt by firing a light 12-bore gun with a spherical bullet is very considerable; as a matter of fact, the recoil is 13 lbs. heavier with a bullet and standard charge of powder than with the standard charge of shot.

A lighter bullet was, therefore, a desideratum, and the "Mead" shell was produced. This consists simply of a hollow spherical core cast in the spherical bullet; it is shown in the adjoining illustration. The hollow core may be filled with an explosive if deemed advisable, and a large charge of powder may be used, and a higher velocity and lower trajectory is obtained than can be got from a spherical bullet of the same size with a shot-gun. The ordinary game shot-gun will not shoot bullets so well as a properly built ball-gun (not rifled); but the accuracy is such that all bullets may be got into a 12-inch circle at forty yards by a good marksman. The charge should not be more than $2\frac{3}{4}$ drams of No. 4 or No. 6-grained powder.

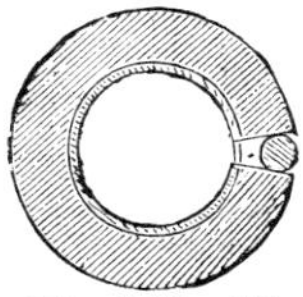
The "Mead" Shell.

For use in shot-guns, special projectiles have been invented which shall fly better than the ordinary spherical ball; some of these have projecting wings wound spirally round the bullet, but the best known are the invention of Dr. Macleod.

In the first model holes are cast in the bullet, and it is asserted that they will be revolved by the air rushing through the eccentric holes as the bullet takes its flight. This heavy bullet—a 12-bore, weighs $\frac{3}{4}$ oz., and is $1\frac{1}{2}$ inches in length—must not be be used in a light gun, and even in a gun of 8 lbs. weight will require an indiarubber wad half an inch in thickness to be placed over the powder before the bullet can be fired without occasioning a painful recoil.

In the last model the loose revolving rudder affixed to the projectile is said to steer it a more sure course. It is open to many of the objections advanced against the first model.

A rifle barrel will not shoot shot well, but the most suitable rifling is

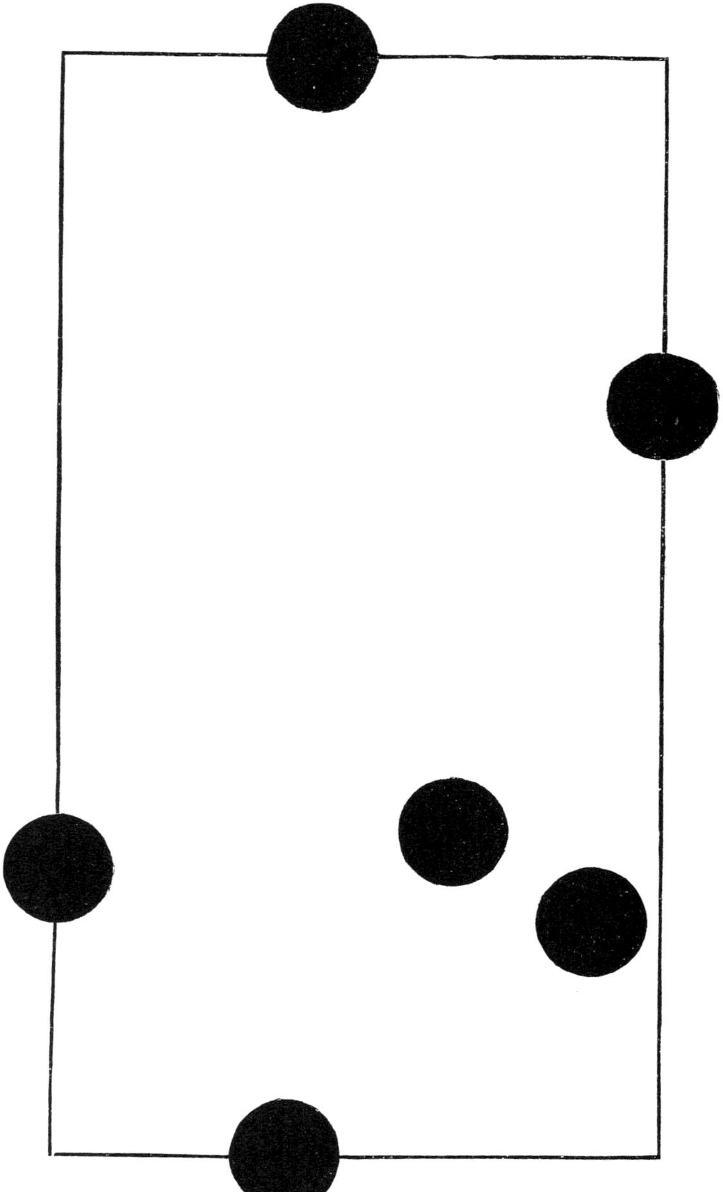

Fac-simile of a Diagram made at 40 Yards range with a light Full Choke-bore Game Gun, 12-gauge, by W. W. Greener.

undoubtedly the Lancaster oval bore, which permits of accuracy with either spherical or conoidal bullets, and a fair average pattern with shot of various sizes.

Gun barrels, when choke-bored, may also be rifled for a few inches at the muzzle, as in the "Murphy" and the "Paradox" guns; and these weapons, although they will not shoot shot as closely and regularly as a true choke-bore, nevertheless perform up to the average, and are accurate with bullets at short ranges. Weapons of this nature should be considered rather as rifles specially constructed for shot than as shot-guns for ball. The Government Proof-House Regulations require that such weapons be proved as rifles—that is to say, tested with ball.

As already stated, the concentrator occasionally acts as a bullet, and the Shrapnell shell may be used as one by turning over the wire spindle.

Choke-bore guns may be used as ball-guns, providing that the bullet to be fired will pass easily through the muzzle, and it may be interesting to sportsmen to know that choke-bore guns shoot ball QUITE AS WELL as guns bored perfect cylinders. Especially is this of interest to those who use but one gun, and have often the chance of a shot or two at big game. Gun-makers and sportsmen alike have been misled by the proof marks; on all choke-bores "Not for ball" has been imprinted.

Another point to be noticed is, that when one barrel be modified choke or cylinder, *it is only necessary to use the one-sized ball*, the larger bored barrel shooting to all intents and purposes as well as the barrel for which the ball is moulded.

Any gun which is safe to use with shot would be quite as safe with ball, provided that ordinary care be taken to see that the ball be not larger than the smallest part of the barrel, and the charge of powder does not exceed the ordinary one used with shot. Further, the ring wads are not at all necessary; one card and one thick felt over the powder, the ball being fixed in either by an ordinary turnover or crimper, will give all that is desired. NEITHER WAD NOR PATCH OVER THE BALL.

The adjoined diagram was made by a full-choke-bored light game gun, charged with $2\frac{3}{4}$ drams powder, bullet 14-bore, and 12-gauge case, distance 40 yards.

RIFLED SHOT-GUNS AND CHOKE-BORED RIFLES.

The endeavours of gunmakers towards producing a good all-round weapon have resulted in such inventions as rifled shot-guns and choke-bored rifles, from which both shot and ball cartridges may be indiscriminately fired. The rifled shot-gun (Fosbery's Patent) has the barrel of an ordinary choke-bored shot-gun rifled for the last few inches of its length, a sharp spiral is adopted, and a grooving of sufficient sharpness to turn a conical-cannelured bullet. Such weapons shoot shot moderately well—much better than the ordinary cylinder shot-gun, but not so closely as the perfectly choked gun—and conical bullets with accuracy to 100 yards.

The oval bore rifle, if the spiral be not too sharp, will throw shot closely and well at ordinary ranges, but a still better weapon than either of these is the Choke-bore rifle, in which a rifle with modern shallow grooving is choke-bored at the muzzle (Greener's Patent) and has a perfectly smooth surface throughout its entire length.

It is made in 20, 16, and 12 gauges, and in all performs equally well. The pattern of the 12-bore with $1\frac{1}{8}$ ounce of shot averages 200 pellets in the 30-inch circle at 40 yards; and with either spherical or conical bullets, or *Mead* shells, it is accurate at all sporting ranges. Whether made as a cylinder, a modified choke, or full choke, the bullet is of the same diameter.

These weapons are treated by the Government Proof Houses as rifles, and are subjected to special tests with ball. They are usually made of light weight—say 7 lbs. 12-bore—and for both bear and boar shooting they offer many advantages, and as a second rifle they fill a need which many a hunter of large game, pioneer, and explorer has often felt. Wherever large game is *occasionally* to be met with, they form the best armament of the sportsman.

It will shoot spherical bullets with a large charge of powder, and will therefore commend itself to many who desire a second weapon for use against buffalo or the *pachydermata*.

RIFLE AND SHOT-GUNS.

The combination of a rifle and shot-gun in one double-barrelled weapon is much esteemed by South African sportsmen. The rifle-barrel, usually the left, may be rifled on any system. Henry rifling is still most in favour at the Cape, and may be of ·450 or ·500 bore; the proper proportions of the two being ·450 rifle-barrel and 16-bore shot-barrel, or ·500 rifle-barrel and 12-bore shot-barrel. These arms are only useful in countries where the kind of game that may be met with cannot be determined beforehand, or for emigrants who cannot afford more than one gun. They have many drawbacks. The weapon is too heavy as a shot-gun, and makes flying shots almost an impossibility. The balance, of course, is bad. As a rifle the weapon is too light, and the recoil with some of the heavier loaded cartridges is considerable.

The heavy rifle-barrel not giving way in the least causes the shot-barrel to become the more easily dented and damaged; a fall to a rifle and shot-gun generally finishes most disastrously for the shot-barrel. The rifle may be chambered for either of the Express cartridges or the long-range No. 2 musket-case; the last-named is most in favour in the Cape Colonies. The Government regulation ammunition may also be employed, but we do not recommend it.

In all cases where practicable, a single rifle and a double shot-gun are far preferable to the rifle and shot-gun.

SECTION VI.—CRITICAL.

THE CHOICE OF A GUN.

THE choice of a gun will be determined, first, by the purpose for which it is to be used; secondly, by the *physique* of the person by whom it will be used.

For general sporting purposes the all-round weapon for use in any country is the 12-gauge gun with 30-inch barrels, weighing 7 lbs.

In Section V. is specified the kind of gun required for any one particular sport, and the following remarks will enable the sportsman to determine what particular gun, of the kind already determined upon, will best suit him. The criticisms and conclusions of the author will be found applicable to guns of all gauges, and for all purposes other than wild-fowling.

The sportsman having decided to purchase a gun, must address himself to either a gunmaker or dealer. And here a word of seasonable advice. In most countries where there exists an importation impost of more than 10 per cent. it will usually be cheaper to purchase of an importer than to import a special gun. In some places, however, the local trader enjoys a practical monopoly, and the prices consequently range very high; it is, therefore, as well to compare the importer's list of prices with that of the home manufacturer, unless there is competition between rival houses. The frauds practised by gunmakers are too numerous and most reprehensible, but it is doubtful if they equal in number or seriousness the tricks of the unscrupulous dealer. The sportsman would therefore do well to go to a dealer of established reputation, or have some assurances of the honest dealing of an unknown man who offers cheap lines.

It must not be thought that every one who offers a Greener gun for sale, or who advertises that he keeps a large stock of Greener's goods, is beyond suspicion. Many of the most unscrupulous make a point of possessing some guns by makers of good reputation, borrowing from the manufacturer's good name somewhat to whiten their own evil notoriety, and thinking that the association of their house with that of a house enjoying an enviable distinction will be a guarantee of their own probity. Choose your dealer with care. The author has appointed agents in various towns, and has satisfied himself of the *bonâ fides* of all to whom he has granted authority to represent him; but many advertise themselves as agents who have no authority so to do.

As with the dealer so with the gunmaker. Go to some maker of established reputation, unless you have very strong reasons for going elsewhere. The author does not for one moment wish it to be thought that good guns are built alone by those makers whose names are household words with shooting men, but it is more probable that the inexperienced gun buyer will be cheated if he purchases a gun with an unknown name, than that he will have purchased the good weapon of the but little-known maker.

Even after deciding upon purchasing of a well-known maker, the choice is practically illimitable, and in determining upon the merits of rival systems the would-be purchaser will find ample opportunities for the exercise of his critical abilities. Should he have set for himself a high standard of ideal excellence, he will find difficulties enough before he has found a gun that will be in every way satisfactory.

As to the material of which the gun barrel is composed, as already stated in the analytical section, there are three species, and of each of these are several varieties.

Of these, the Belgian barrels will be condemned. Made of much softer metal than English Damascus barrels, they will not stand full choke-boring, and so soft and yielding are they that they bulge at the choke when fired, and cannot be made to retain their shooting qualities. These remarks apply to all qualities of Belgian Damascus barrels.

Of English twist barrels there are two types—the machine-made, which have a rather coarse appearance, and often present irregularities of the figure; and the hand-welded barrels, which contain a larger percentage of steel, and are now very difficult to obtain.

For general purposes the machine-made barrel will be found serviceable and safe, and even if it have a few specks, it may be taken, for such small stains are of no practical importance or real detriment to the gun, and it is almost impossible to have a barrel of this description entirely free from them.

The best hand-made English Damascus or laminated steel barrels are made at Birmingham only, and are without a doubt unrivalled the world over. For beauty, durability, and reliability they are far ahead of every other description of barrel whatsoever.

As they are closely imitated by the Belgian welders, purchasers of guns who feel in doubt as to whether or not the barrels of the gun offered them are of English or foreign manufacture, should ask for a written guarantee that the barrels are English-made and of English metal; and as the dealer, if an expert in technical matters, as he should be, cannot be deceived by Belgian imitations, he can have no hesitation in giving the written guarantee required; but it must be remembered that a reference to the *proof mark* will not reveal the gun's origin. Belgian barrels have been worked up by incompetent English gunmakers to the great detriment of the reputation of laminated steel and Damascus barrels generally, and this practice may yet long continue.

The greatest blame attaches to those London gunmakers who have long used, and continue to use, an inferior Belgian imitation of a genuine English product, and which they know, and have long known, to be inferior.

Steel barrels, even of the best quality, will not stand heavier charges than the best barrels of English laminated steel, nor do they show any superiority over best Damascus or laminated steel barrels in any way. They offer certain advantages to gunmakers which it is not the object of this book to disclose.

Steel barrels, even of the best quality, may be rendered unsafe by brazing, but there is a way of testing these barrels after they have been brazed which can be absolutely relied upon to indicate their soundness, or reveal the nature of their deterioration.

A gun with steel barrels should only be purchased of a reliable maker or dealer, and if a certain brand of steel is required, a written guarantee should be demanded, as the steel itself does not show any mark by which the true "Whitworth" or other brand may be known.

It is necessary that the gun be safe when fired with any charge of any explosive that can be got into the cartridge case for which the gun is chambered. It must be safe also from accidental discharge. It must have a breech mechanism that will wear, and be easily manipulated, easily loaded, and as easy to unload. It must be pleasing to the eye and fit like a glove, kill anything at every reasonable range, and the cost must be moderate. In fact, every good quality is expected without the accompanying drawbacks, and a perfect weapon is expected because a gun is something which can be built to order, and need not be purchased ready-made like a horse. The safety of the weapon with any charge should be guaranteed by the Government Proof House, an institution for which the gunmaker is heavily taxed. As a matter of fact the Government Proof House test has long been useless so far as any guarantee of the gun's safety is concerned. Guns and barrels brought direct from the proof house have burst in the hands of gunmakers' assistants, or whilst otherwise undergoing such tests as the maker has thought needful to determine the safety of his weapons. Reform has long been needed, and after long agitation certain alterations have been proposed, which, when practised, will be some guarantee to the gunmaker that the weapon proved is a sound one.

All gunmakers of reputation test their guns with heavy sporting charges of *sporting* gunpowder, as well as making sufficient tests of the weapon's shooting powers. Until the Government proof test is an all-sufficient one this will be absolutely necessary. The proof-mark, therefore, will not be taken by the buyer as of any practical value, although he will, of course, reject at once any weapon not possessing it. The guarantee of the gun's safety in this respect must be the name it bears.

Any gun upon any system of lock mechanism will be safe from accidental discharge, providing the lock mechanism has been made and fitted by an intelligent and experienced man, and the work done carefully with the best material. The price, therefore, of the weapon will be a guarantee of the safety of the weapon on this point. In the choice of breech mechanism a top lever will certainly be decided upon. Side and bottom levers will not be chosen, save for very special purposes or to accommodate a sportsman habituated to this form of lever from long use. The top lever so conveniently placed enables even the most determined adherent to any other form of lever quickly and comfortably to manipulate the breech mechanism. As to holding-down bolts and cross or grip bolts, the number and position

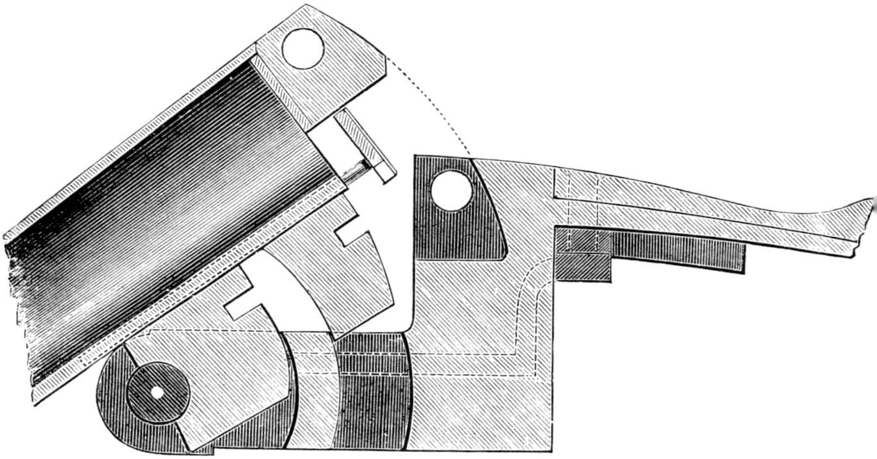

The Working and Bearing Parts of a Hammerless Breech-Action.

of the barrel lugs, the sportsman will stipulate for two binding and well-fitted bottom lumps or lugs, substantial in size, and the back lump so shaped as to correspond with the *circular* front of the slot in which it is embedded, and so well fit that when the gun is closed and fired this back portion of the lump shall take part of the strain as well as the hinge pin. This method of fitting the barrels to the breech mechanism is technically termed "jointing on the circle," and the accompanying illustration will convey a better idea of its nature than any verbal description.

The need of a larger number of barrel-lugs situated under the barrels has never been made manifest, although some gunmakers put four instead of two. Certain descriptions of guns, by reason of the arrangement of other mechanism, may, however, cause the number and position of the bottom barrel lumps to be other than these described as generally advisable. It is the mere multiplication of these lumps for the sake of intricacy that the author wishes to warn sportsmen against.

Guns very carefully built and carefully used with moderate charges have withstood even continual wear for years without other bolts than the ordinary double holding-down, or, as it is sometimes called, "Purdey" bolt; but, generally speaking, it is advisable to have a sound top connection between the barrels and the breech-action. It is always safer to have such a fitting, and in some cases it is a positive necessity. The top connection, to be of real service, must be a good one ; not a sham, not a make-believe "giant," "hold-fast," or "triple-wedge-fast grip."

The common top connection is a "doll's-head" fitting more or less

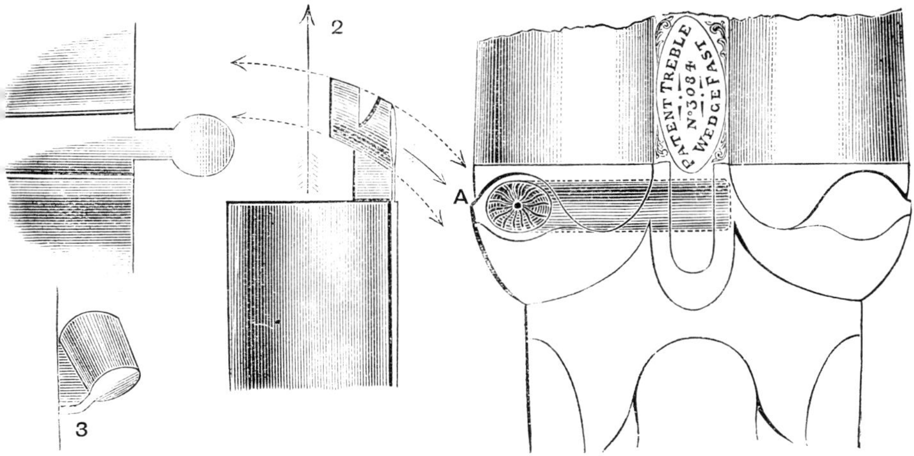

Top-connections between Barrels and Breech-Action Bodies.

badly into an ill-shaped hole, hollowed out in the standing breech to receive it. (See No. 3 in the accompanying illustration.) This useless encumbrance is affixed to guns to appease the ever-increasing demand for an efficient top-connection which the success of the author's and the Westley Richards' mechanisms has produced. No. 2 of the illustration shows the Westley Richards'—the best—form of "doll's head." The strain upon firing is in the direction of the arrow, but at the same time the expansion of the barrels conveys to the extension a tendency to move in the direction of the shorter arrow (A) ; the bolt into this top extension in some measure prevents movement in the direction of A, but much depends upon the manner in which the bolt is fitted. At the best this contrivance is inferior to that mechanism invented by the author in 1873, and now is well known as "Greener's Treble Wedge-Fast." The principal feature in this mechanism is the cross-bolt working transversely in the standing breech, and wedging the barrels to the breech-action body at the top. No. 1 in the illustration shows this bolt. It is worked by an arm on the top lever, and adds immensely, not only to the strength and stability of the breech mechanism, but will cause the gun to withstand successfully continuous wear.

The following is the best proof that can be given of the necessity of a top connection, and the advantage of the "Greener cross-bolt."

After the close of the *Field* trial of explosives in 1878, the manager (the editor of the *Field*) made some experiments as to the breaking strain of powders on gun actions. The author constructed a gun in such a manner that it could be used either with or without his "cross-bolt." To this gun a small arrangement was fixed, as in the adjoining illustration.

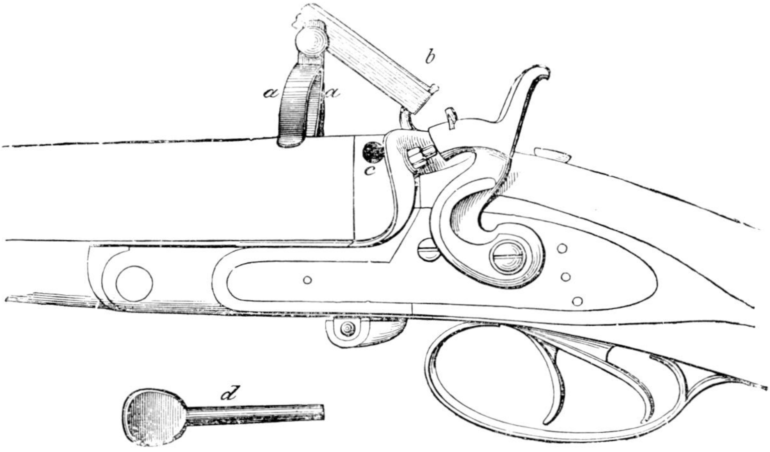

Experimental Breech-Action.

Between the hook (*b*) on the break-off and the screw-clip (*a a*) attached to the barrels a piece of silver paper can be strained, so that when any separation between the barrels and the breech-action takes place, during an explosion, the paper breaks. Both barrels are loaded equally, after which one is fired with the bolt in, and then, supposing no breakage occurs, the bolt is removed and the other barrel is discharged.

"Experimenting in this way," writes the editor of the *Field*, "we found that in Mr. Greener's action no breakage occurred; with the bolt in position, the paper remained intact up to the last." With sixty-five grains of Schultze powder and with seven drams of Curtis and Harvey's No. 3, there was a slight breakage of the paper (the top bolt being out), and with 75 Schultze powder, "there was not only complete breakage of paper, but such a permanent opening of the breech of the gun as to stop the experiment."

Mr. J. H. Walsh, in his work "The Modern Sportsman's Gun and Rifle," writes in flattering terms of this action, whose advantages he was one of the first to demonstrate, and it is now even acknowledged by contemporary gunmakers to possess great merit.

It is far superior in strength and lasting power to the double-grip action. A double 4-bore and a double 8-bore were made on this, the top cross-bolt, principle, in 1874, for Mr. G. P. Sanderson, superintendent of the Government Elephant Keddahs, Decca. They have been in continual use ever since, firing 2-oz. bullets with 12 drams and 4-oz. bullets with 16 drams of powder, "hundreds of times," and to quote Mr. Sanderson, "the breech-actions are as sound and close as when they left the factory nearly ten years ago." These rifles are still in use, and doing excellent service.

A rifle-maker of good reputation essayed some time ago to make one of their large-gauge elephant rifles on the double-grip plan, using also the "doll's-head" extension as a top connection. Three breech-actions were successively fitted, and all were strained and broken in the preliminary trials, and the prospect of getting one to stand the heavy sporting charges was so remote that the construction of the rifle upon that plan had to be abandoned. The author can vouch for the truth of this statement.

A Greener Treble Wedge-Fast Gun may be fired hundreds of thousands of times without any repairs being required. To the author's knowledge the gun he supplied to Dr. Carver in 1878 was fired 130,000 times without any repairs being done or required; and the gun is still in use, and has had possibly thousands of shots fired from it since the author last saw it.

The sportsman will choose a gun with this top cross-bolt, providing he wants a gun that will fire any charge without the breech-action giving way, providing he requires a light gun—a gun for very heavy charges—a gun for hard wear—a gun that will last him and always be satisfactory.

The tendency now is to build guns with shorter breech-actions than formerly, and consequently the strain upon the breech mechanism is greatly increased, and renders the use of the top cross-bolt still more imperative.

With a top cross-bolt both barrels may be fired together without the least possibility of the breech-action giving way, and as this is a contingency that may at any moment accidentally arise, it is as well to be furnished with a gun that will stand both barrels being fired together.

HAMMER OR HAMMERLESS.

The introduction of any important novelty to the shooting world is the signal for the introducer to be bitterly assailed; and discussions are promoted in which arguments are advanced, which not only have no bearing upon the subject, but divert into side issues the attention of sportsmen, and so keep the main question unsettled. It is this way with the "Choke" introduced in 1874, and with hammerless guns.

The leaders of the fashion in shooting circles and the accepted authorities upon gunnery beg to differ, and sportsmen are left in doubt as to the merits or disadvantages of any vaunted novelty. The would-be gun critics are too often better judges of dogs than of guns, and with no accepted standard of what a gun should be or do the dictum of an ignoramus will occasionally pass as a pearl of wisdom.

One shooting newspaper has quite recently discovered that choke-bore guns are all a mistake, and advances ten years after their introduction the very same arguments which were advanced and disproved in the last decade. So is it with hammerless guns. We find Mr. Walsh, the well-known editor of the *Field*, writing that "the hammerless gun is the superior in point of safety and efficiency. . . . The hammerless gun is, I think, to be preferred." On the other hand, in the Badminton Library, we find the assembled and aggregate wisdom of the leading shooting men of Great Britain, as written by their deputy, to be: "They (hammerless guns) are not, as at first maintained, quicker and easier to load than are hammer guns, and are liable to damage;" and a page is devoted to prove that hammerless guns are not any improvement upon hammered guns. The last sentence, however, commences, "As we, however, believe that hammerless are unquestionably the guns of the future, we have depicted," &c. Even the most prejudiced against hammerless guns cannot *convince* themselves of the *equality* even of the hammered gun.

There is often a cause for this prejudice. The sportsman with a pair of old guns, and on the look-out for new ones, is absolutely without prejudice. He goes to his gunmaker, a most worthy and honest man doubtless, who has served him and his friends well for years, and he accepts the notions of his gunmaker. The gunmaker may have £15,000 worth of hammered guns, old stock, and no interest in any good new hammerless gun; or he has not invented, or introduced, or recommended choke-boring, and he is hearty in his denunciations of the "fads of Brummagem." This opinion is passed on. Our dog-breeders, editors, and authoritative sportsmen call on this gunmaker and find it easier to pass on his opinion than form one for themselves.

But the gunmaker's stock gets older and mustier, and as its value decreases his denunciations become more fierce and worthless, and this is how it comes that old diatribes are reiterated year after year, when the value of the invention itself has long been proved beyond all question.

The greater number of gunmakers more or less readily take up a novelty, making joint or several effort to break through any monoply, and when they can make or sell any other person's invention with impunity, do not bate one jot from the inventor himself in praise of the novelty. The first endeavour is to secure a monopoly, the next to realise their old stock, loudly declaiming against the novelty whilst they do so; then push *their* modified form of the novelty. If the newspaper critic is inspired by this gunmaker, he writes for hammerless guns; if by "Old Stock," the gun-seller, against them. If he uses the guns, he writes in praise of the new weapons.

Having detailed so much of the *modus operandi* followed in producing criticism, the author will proceed to state concisely why hammerless guns are to be preferred to hammered guns.

1st. They are safer.
2nd. They are quicker.
3rd. *Cæteris paribus*, they are stronger.
4th. They are less liable to damage.
5th. They are more reliable.
6th. They are less complex.
7th. They are more handy.

They *may be* more durable, more economical, and more beautiful.

The hammerless gun of inferior quality is as dangerous as the inferior hammered gun, if not more so; and in the choice of a hammerless gun the sportsman will be guided by, 1st, the simplicity of the mechanism, 2nd, the efficiency of the mechanical parts introduced to effect that hitherto performed by hand.

It has already been shown in Section III. that a hammerless breech-loader can be constructed with fewer parts than a hammered breech-loader. The sportsman must choose a gun so constructed. If a special lever is desired, as under, side, or the double-grip lever, good guns have been devised on each of these plans. There is the "Greener" Club, with under lever; Grant's, with side lever; and Allport's, with the double-grip lever. The top lever will doubtless have the preference, and here there is abundant choice. All guns cocking by means of mechanism geared or in any way connected with or dependent upon the motion of the top breech-action lever for effecting the cocking of the gun, will be at once rejected. Because it is liable to miss-fires if the lever does not snap "home," and because it is *generally* fitted with weak mainsprings, and often requires great force to open.

Of those guns cocked by the falling of the barrels or closing of the gun will be rejected all that, 1st, do not permit of the barrels being placed readily upon the stock; 2nd, that may be wrongly put together and so cause a breakage; 3rd, all that require a jerk to open or shut. Because all such guns will be an annoyance to the owner, and in the case of the last objection will cause undue wear at the hinge joint and need early repairs.

Of guns which work easily and pleasantly, and have none of the above objections, will be rejected all that have the mechanism so arranged as not to be readily accessible to determine whether or not any part be fractured in case the gun refuses to function, or whether or not a lubricant should be applied. There are several guns which will fulfil every requirement of the sportsman so far as mechanism goes. The Anson and Deeley is a simple mechanism, which has only a few of the objections the author has advanced.

Greener's *Facile Princeps* has merits not possessed by the Anson and Deeley, and it is different to it in principle. The sportsman will make himself acquainted with two or more rival systems before purchasing, and he will learn sufficient concerning the mechanism of the gun he makes his own to inspire him with complete confidence as to its safety and his ability to use it.

And one word of warning here as to the "safety" of hammerless guns. The safety is an adjunct, not a necessity, and if badly constructed or faulty in principle it may be an instrument productive of danger and fraught with annoyance. There are several simple and efficient safeties open for any gunmaker to make use of, and it will usually rest with him to fit such a one as will best suit the particular gun he is constructing.

The author recommends the hammerless guns strongly, he has a large stock of hammer guns unsold, and each hammerless gun he makes costs him more than the same grade of hammered guns; but he is so fully convinced of the merits of the hammerless gun when properly made, that he has always been one of its strongest advocates, and has made more than any other gunmaker. Of the guns he has made fully 99 per cent. have been fitted with trigger-bolting safes not automatic in their action, and these have given complete satisfaction, not a single complaint having been made concerning them.

The self-cocking of the locks is in itself an immense advantage, and the self-ejecting of the fired case, by reducing the number of movements to be made by the hands when loading, is decidedly worthy of every support. The self-ejecting mechanism employed by the author, and described and illustrated in Section II., is of the simplest kind, and invariably does the work required of it well and with certainty. It has been well tried in every quarter of the globe, and is recommended by all who have made use of it. It stands at the present time at the head of sporting guns, the nearest to perfection.

The following opinions or excerpts from lengthy arguments in favour of hammerless and ejecting hammerless guns shall close these remarks on the choice of a hammerless gun:—

"The absence of hammers makes the gun very convenient, especially for covert shooting, to which I cannot speak too highly of its superiority combined with safety and ease of manipulation. "GERALD L. GOODLAKE, Col."

"The hammerless guns you made for me about four years ago have stood remarkably well; they have never been out of order; the locks have never been taken off; neither has the safety-bolt been taken out or cleaned. "GRANARD."

"I like the gun very much indeed, and find it very handy. The action, too, is very neat, and an improvement on the older pattern of hammerless."

"I do not see that there is anything in the ejecting action which will make them wear out sooner than other guns; at any rate, mine work as smoothly and perfectly now as they did the first day I used them."

"I received the ejector last autumn. I returned it three weeks since. It was never in that time out of my possession, was never out of gear in any way. It had done lots of work before it came into my hands, and is as ready to go through as much more. . . . Its three advantages appear to me to be *celerity*, *comfort*, and *economy*. An ejector gun *will* do the work of *two* guns."

"The Ejector Gun is a step in the right direction; it throws out the empty cases perfectly. This ejecting is a great convenience in a hot corner."

"With reference to the safety of hammerless guns, I agree with you that a safety bolt, to be of any value, ought to be reliable, and to illustrate my meaning, the following may interest your readers:—When grouse driving on the Berwyn Mountains, in Montgomeryshire, some ten days since, and using a Greener ejecting gun, with safety bolt, the rain commenced descending in torrents, and, as a fog seemed imminent, the order was given for home. Instead of following the downward track adopted by the keepers and beaters, I decided on a shorter line of country, and decided to make my way along the face of a steep hill, scantily covered with fern. Placing my gun, which was loaded, at 'safe,' I made the attempt, and got on fairly well for a hundred yards, when I suddenly slipped and began rapidly to descend. After going some forty feet, and finding the pace increasing, I was forced to let my gun go. Slipping and swinging round, presenting its stock and muzzle alternately at my head, it shot rapidly down the hill and disappeared over the cliff, towards which I unwillingly followed. My sensation at that moment I keep to myself. Luckily some friendly ferns checked my pace, and I brought up a few yards from the edge. Regaining my feet, I cautiously proceeded till I got on a sheep track, and succeeded by the aid of a boulder in gaining such a foothold as to enable me to approach the edge and attract the attention of the men, then hundreds of feet below me. Indicating that I had lost my gun, one of them with great difficulty climbed up the face of the hill, and after some time uttered a shout. Then, far below me, and embedded half-way up the barrels, with the stock sticking straight up, I perceived the gun. A mossy spring between two rocks had received it in its fall; a couple of yards to the right or left, and it would have been smashed to atoms. Twenty minutes later it was restored to me, the barrels plugged up for some inches, but apparently having received no external injury, save a few scratches, and a piece chipped off the heelplate. The trigger guard was, however, a study; bits of fern and rushes were twisted round the triggers, which caught in everything in the downward course; but the safety bolt had done its work, and the cartridges were intact. While I write there hangs above me an old and valued servant, a Greener gun with rebounding locks. Nearly 100,000 shots have been fired out of that gun; had it, however, been with me on the hills that day, a different sequel might have been told. I always considered hammer guns with rebounding locks required care, not only in crossing fences and in covert, and from the liability to explode when dropped, but from the tendency of the hammers to catch in buttonholes and watch-chains, as has frequently happened to me. Sportsmen have every reason to be thankful that science has invented such a boon as hammerless guns with reliable safety bolts—in my opinion the safest and pleasantest guns any one can desire—provided gentlemen recognise the fact that a cheap gun on that principle is one of the most dangerous things out, and, when they decide to go in for a hammerless gun, select a first-class gunmaker for the purpose."

For speed the self-acting Ejector gun is ahead of all magazine or repeating shot-guns, and not only can it be fixed more quickly, but it is free from any liability to jam when rapidly manipulated, whilst the repeating mechanism of shot-guns is more prone to "jam" than the mechanism of a rifle, owing to the use of paper cartridge cases, turned down to various lengths.

THE BEST GUN.—THE QUESTION OF PRICE.

The sportsman, now decided upon possessing a hammerless gun with self-acting extractors, as the very best with which he can furnish himself, considers the price. Why should hammerless guns cost so much more than hammered guns? and why should there be so great a difference in the price of hammered breech-loaders? A double-barrelled, central fire, 12-gauge breech-loader, proved, and a complete, usable weapon, is sold wholesale, at the present time, at twenty shillings. At that price it is at present a marketable commodity, and the tendency is downwards. A best hammered gun, 12-gauge, proved, a complete, usable weapon, is to be purchased at sixty guineas, and will not be sold for less. Is the £62, the difference between the two, solely for the maker's name engraved between the barrels? If not, where is the difference to be seen?

To decide this, let a best gun be described, and the details of its manufacture be somewhat minutely gone into. The barrel is made of a splendid quality of metal, costing a fabulous sum per ton. In constructing a pair of tubes 18 lbs. will be used, and the metal will be worked at a red heat, and from its multitudinous welds, as well as from its inherent quality, will gain strength, *i.e.*, tenacity. This 18 lbs. of iron, by hand-labour, has to be reduced to *three*, and the superfluous metal must be removed just as and when required, or the barrel will be worthless. Through twenty-six succeeding processes these tubes have to be handled, worked, and treated with care, and may, after receiving every attention human skill can devise, prove defective at the last moment, and instead of figuring as the best gun, be relegated to the waste heap—value 3d.

The workmen in every branch of the gun trade are divided into classes. The careful workman, mindful not only of his work upon the gun, but cognisant and careful in his treatment of the work of those who have gone before him—skilled, and able to do what is required and expected of him—is a *rara avis*, who can command a high wage. A staff of such men *must* be procured if the best work possible is to be obtained; and they must not only be kept fully employed, but employed upon such work as they can take an interest and pride in. To produce a best gun, not only must every man be able, but inclined, to do his best; and above all, there must be the guiding mind, intent upon the fashioning of a weapon to its ideal.

The best gun must be tried in various stages, and must pass in each before proceeding to a succeeding stage; hence time as well as money is requisite to its production. The well-finished gun is one in which every portion is accurately shaped, rightly placed, perfectly adjusted, and with that "finish" which skill and practice alone can give. The elaborate ornamentation, either by engraving or otherwise, will not make a gun well finished; nor is such ornamentation of such use as finish. A gun made and finished in the best manner will stand more hard wear than any ordinary gun, even if the principle upon which the commoner gun is constructed be superior to that of the best gun. Common guns always give way first in the small details: a pin works loose or breaks, and as soon as it is replaced in one place it gives way in another, whereas a best gun, like "The One-Hoss Shay," breaks up altogether *when* it does go. If you want to enjoy sport, have a best gun; if you must shoot, and do not mind when, take an ordinary gun; it will fail you when you have a first-rate chance to bag game, but the best gun is always good alike, and thirty seasons' hard wear will leave it as good as new for all practical purposes.

HOW TO DETECT A SPURIOUS GUN.

A man may be a very good shot, and have used a gun from childhood, and yet be practically unable to appraise a gun. This is somewhat remarkable, seeing the numerous instructions which have been published to those about to purchase guns, and taking into consideration that the sporting newspapers have generally a column open to the discussion of this subject. It very possibly arises from the wilful misrepresentation made in advertisements, the bias with which most contributions respecting the merits of guns seems to have been penned, and the increasing prevalence of trade puffs. Strange to say, those keen people inhabiting the United States of America appear less able to discriminate between a good and a bad gun than any civilised people. No nation possesses better sportsmen nor a greater number in proportion to its population, yet we find the Continental rubbish-factors exporting in great numbers such weapons to the United States, that he cannot for shame offer for sale in his own country or any British colony. The Boers are a race of sportsmen, but it is of no use offering them rubbishing weapons at *any* price, and the author can hardly believe that the astute American will sacrifice everything to cheapness. There is certainly the fact that the American salesmen are without equal, and have such powers of persuasion, that one is half inclined to believe that the American rifle has never had its equal; but even the ability of the salesman could not overcome the repugnance of the buyer to the rattle-trap designated by the Suhl or Liège maker as "export guns," providing the would-be purchaser could or would discriminate between a serviceable and an unserviceable weapon. In the United States there are two classes of guns made. The machine-made trade gun, the sale of which is vigorously pushed at every opportunity. The better-class gun, made by some American-born or emigrant gunsmith, whose production is limited and sales unimportant. An American gun, at about three times the price of the American machine-made gun, will be a superior weapon in every way to the machine-made gun, but be sure that it is of American make, for *imported* guns are sold as of any make, just as there is a demand. Of imported guns there are three classes—the real trade gun, rubbish; the legitimate trade gun; English or foreign guns, made sound and well by a responsible maker, who will put his own name upon them, and give as good quality as the price given by the importer will allow; the fine gun, the *bonâ fide* production of an English maker of reputation, and imported to special order, or for sale only by the special agent of the maker in question, or some honest and enterprising dealer. In America, however, dealers are very loth to keep in stock the fine guns of any maker. In England, on the Continent, especially in France, Germany, Austria, Russia, and Italy, where the sportsmen are more discriminating and exacting, there is always a choice of twenty different *grades* of guns, and—especially in France and Germany—the sportsman can appraise the additional amount spent in bettering the quality of the weapon. The American, and very many colonial sportsmen, cannot or will not discriminate between the first and second classes, and are slow even to see the difference between the second and third. Now, nothing should be more easy than to distinguish the good gun from rubbish; the third from the first of the classes before referred to.

The gun which costs 60 guineas might not be easily distinguished from one which costs but 59; but in the "trade gun," the price of which has been gradually falling, and has now dropped to 20 shillings, the drop of *sixpence*—to gain which the importers will ransack the manufacturing centres of Europe—is missed as the nose from a man's face.

Look at the appended illustration: the cheapest gun is here depicted; it may be known by having—1st, all the parts which should be square and flat, rounded; 2nd, all the parts—as the barrels—which should be round, a series of flats; 3rd, hammers which are odd, and which stand when both are at half-cock as though one were at full, and, when both are "down," one rests on the nipple, but the other will not reach it; 4th, one lock won't "speak," the other roars; 5th, one striker sticks out and upwards, the other is pitched as though the breast—not the head—of the hammer were to strike it; 6th, the rib is not straight, and is very much more on one barrel than on the other—the barrels are neither straight nor round, and are generally thicker on one side than the other; 7th, the extractor has a crooked leg, and when the gun is opened, it sticks out as though pleased to escape from its ill-shaped recess—on closing the gun, its contortions are astonishing; 8th, the barrels are bright inside, but it is not the brightness of a silvered mirror, rather the brightness of a leaden bullet; 9th, there is no close fitting of any part: the action body is barely touched by the barrels, the holding-down bolt is a crooked article in a crooked hole, the fore-end will drop from the gun when it is fired, or will want all your strength to get it off, and the "wide joint" may be seen wherever two pieces come together; 10th, the engraving is a series of ill-shapen, deeply-cut furrows, cross-harrowed with meaningless scratches; 11th, the balance is bad, and the gun heavy; 12th, the stock worse than that of an army musket, having traces of "file-teeth," and exhibiting that rough open grain inseparable from spongy wood, and which the oily gloss cannot hide;

13th, the butt-plate, an ornamental sporting or other design made of stamped rubber.

Such is the "export gun." If its user survives ten shots, the gun will not. On trial it may fail to go off; the striker is too short, or does not strike centrally; this is rectified; then it will be found that the other striker is too long, and, after the gun has been fired, it will not open: this is altered. The mainspring is so poor its elasticity has departed, and miss-fires ensue; new mainspring fitted: this is too strong for the lock, which is only of soft iron, so the tumbler gives way; steel tumbler fit: the scear, being iron, has worn away in only trying the lock, and fails to keep lock at cock, so the gun goes off unawares; complete new lock-work fitted: hammer drops off, triggers jam, and screws drop out in an unaccountable manner. The gun is thoroughly overhauled, is kept a month at the smith's: at first shot barrels drop asunder, owing to having been soldered together with sal-ammoniac, which, from its chemical action, destroys barrels and solder. Thus the cheap gun costs more in repairs in one season than a good gun would want in twenty, and is a standing annoyance to its owner. The gun of slightly better class will *look* much the same, but the locks should be of steel, and the action fitting better. Twist barrels are a step higher, and triggers pivoted in an iron "box" are higher still. Then come grades with better fitting, and traces of some care having been used in putting the strikers in centrally, in getting the hammers to match, in having

The "Export Gun."

the rib midway between the barrels. When Damascus barrels are used, the gun is up in price, and the weapon reaching a serviceable standard. Next, the barrels are straight, the stock harder and more shapable, the lines cut into the iron can be seen to follow some design—fugitive and inappropriate, it may be, but still a design. With smoothly working locks, better balanced guns, two iron Damascus barrels, usable pull off, and a well-fit action, we are rapidly approaching a grade that may be serviceable, if not high-class. When, instead of a rubber-stamped butt or heelplate, we have an ebonite or horn *hand-chequered* one, we have reached the first grade of the artist workman, and not the turning-out machine. We find in the better grades a smoothness and flatness of the lock-plates that is easily noticeable; and, as the inside of the plate is square and flat too, the lock is cocked with an easy movement, and uniform increase of pressure. Not only do the hammers match and stand alike, but nipples, triggers, and screws fit closely and tightly; and in the still higher grades every pin will be found to fit accurately, to have its slit running in a preconceived direction, and *every part*, when inspected, will be found to have had some attention paid to it, to make it as perfect as the worker's idea of it had determined. In examining a fine gun, even if it be as heavy as that of the "trade gun," it will be found to handle "like a thing of life" when compared with its "export" competitor; the bottom rib will be found as accurately shaped, as small, and as carefully put on, as though that were the rib which would receive every scrutiny; and even the butt-plate screws—which to the well-glued heelplate are of very little service—will be found to be as well-shaped, slit, and accurately fitted as if the whole reputation of the gun and its maker were staked upon those pins alone. So must it be. Unless attention be given to *every* piece, no matter how seemingly unimportant, the gun is not well made, and may fail just where least expected.

From the first conception of the gun to the last stroke of the buffstick, there must be paramount care in the choice and fashioning of the material, and the right relation in size and position of every piece to each other and to all.

There is probably no gun without its faults of construction, but in a gun of the first quality they should be known only to the maker, and such as *he* cannot remedy nor others detect.

Then, just so much as is the talent of the maker superior or inferior to that of his competitors, will his gun be superior or inferior to their productions.

The machine-made and machine-finished gun may be distinguished, 1st, by its rough "metallic" appearance; 2nd. The heads of the screws will be found to be rounded down, not filed off flush with lock-plates, breech action, butt-plate, &c. The slits of the screws lying at different angles; 3rd. Inequalities in barrels and rib; 4th. The stock shaped like an army musket. In hand-finished machine-made weapons many of these irregularities will not be noticed, but as long as the gun is interchangeable, many niceties in construction and finish will have to be sacrificed. The shaping and trueing of the barrel exteriorly and of the lay and evenness of the top rib cannot be done by machinery so well as by hand, and although machinery is an excellent aid to the gunmaker, by relieving his skilled workmen in the removal of surplus metal, it can rarely be utilised in fitting, and it is the endeavour to leave the various parts "from the machine" that results in so much bad work being found upon "interchangeable" guns. As a general rule, the further removed a machine-made gun, is from interchangeability the greater its worth. All the machine-made guns the author has seen, whether of English, American, or Continental make, have such faults as to interfere with their utility, and cause them to be ranked as weapons inferior to hand-made weapons at the same or even less cost.

The spurious gun may be either a gun represented as being of a quality it is not, or as the production of a maker other than the real one. After taking all into consideration, it is the first class which is the most dangerous to the unwary buyer. The vapid platitudes of the salesman spread a glamour over the transaction, and the sportsman purchases a gun which will trouble him more and more as he gets to know it. Against the purchase of this class of gun the sportsman must always be on his guard.

The second class of gun is simply a forgery. Belgian guns are sent to England to be proved, or the English proof marks are imitated; "English fine twist" is engraved upon the rib, or *any* maker's name is put on to the order of the importer.

Some makers do not scruple to state in their lists that they will put upon their productions "made in London, or in Eibar, or in Brescia," or in any other town whose manufactures have a better reputation than their own. *Never* buy a gun without the maker's name upon it.

All the leading makers or their retailers now advertise, so that the exact name of the maker wished is easily obtained; see that the gun bears this name, and rightly spelled, for the change of a letter is often made, the maker of the forgery thereby thinking that his liability is lessened, and foreign forgers make dreadful havoc with English names, whereas probably no careful maker has ever turned out a gun wrongly or incorrectly named, so far as *his* name goes. As to the more general forgeries, they will be found to be changes rung upon the name of a maker of reputation. No one would forge "Smith" or "Jones," and happy the gunmakers who possess such names; but names as "Greener" will be spelled "Greenen," "Purdey" as "Purdy," "W. C. Scott & Son" as "J. N. Scotts Son," whilst of the imitations of "Westley Richards" the name is legion. The alteration in the initials, or the Christian name, or the address is more frequent, and all "Horace Greener," "Albert Greener," J. H., W. H., A. H., and other H. Greener guns are practically forgeries. From the affluent position most of these dealers and getters up of spurious guns enjoy, makers of reputation prefer to suffer than engage in what they know must be a disagreeable and very probably a most disastrous prosecution. The author believes that he alone has instituted criminal proceedings for this species of forgery; the result being the imprisonment of the offender. And although the method of procedure is distasteful and expensive, the author appeals to those who have been deluded into the purchase of a forged Greener gun to communicate with him at once, in order that an effort may be made to stop this nefarious trade.

In Great Britain, under the New Merchandise Marks Act, makers of spurious guns may now be prosecuted, and the sooner the chief clauses of this Act are made International law the better it will be for foreign sportsmen. There can be no doubt but this Act will practically revolutionise the gun trade if its provisions are rigidly enforced, and as guns are not generally manufactured in London, it will doubtless be the London dealers who will first suffer, whilst the reputation of genuine Birmingham and London made guns will be enhanced.

As no two persons are exactly alike, each person to be perfectly suited with a gun will require something different to that which will suit any one else; but as perfection can never be obtained, the nearest to it must be found. This medium will be found nearer to perfection than to the "machine-made" gun, which is designed as if all who are to use it were made in one mould as is the weapon. The gun must "suit" the person who is to use it in the following particulars: weight, amount of choke, and the shape of the stock.

Many sportsmen, and not a few good trap shots, sadly over-weight themselves by making use of needlessly heavy guns. The gun "put up" at a mark for trial does not seem heavy, but after carrying it a few hours, or sitting by the side of it for half a day in an open boat, or waiting about the match ground any length of time, will cause fatigue, and the gun will not be "put up" as it was when the body had no such need of rest. The lighter the gun the longer the tired muscles retain control over its position; therefore, carry the lightest gun the nature of the sport will permit. The question of recoil is associated with that of weight; with a light gun the recoil is more appreciable than with a heavy one. The recoil must ever be unpleasant with the standard charge for the gun. Heavier charges may be used upon special occasions, but as these circumstances will be exceptional, it will not invalidate the rule. To habitually use a charge which produces unpleasant recoil is most foolish, and as neither the gun nor the shooter can long withstand the unnecessary strain, it will increase the bag, and save money to reduce the charge.

The amount of choke necessary will be regulated both by the nature of the sport for which the gun is chiefly intended and the skill of the shooter. A trap shooter placed at 24 yards must change his gun when the handicapper puts him back to 28. In deciding as to the amount of choke required, it must be borne in mind that the pattern made at any given distance does not fairly represent the position of the charge at any given moment. The pellets of the charge issue from the muzzle as compactly as a ball, then having individual velocities—some go ahead, others lag behind, so that, roughly speaking, there is a distance of twelve feet between the first and last pellet when the bulk of the pellets arrive at 40 yards.

For trap shooting a gun is required to shoot as closely as possible at the trap. The bird must be shot at quickly, and the nearer to the trap it is grassed the better. Naturally, the pigeon-shooter requires as large a killing circle as is compatible with a close pattern. He requires the pattern to be equally spread over the "killing circle" to have the greatest velocity, and the pellets to keep together as much as possible; but of greater importance than all these, it is required that the gun always perform alike. Uniformity in shooting is a quality only found in the best of guns, and even cylinders, unless most carefully finished, will make occasional bad shots, any one of which would allow of the pigeon escaping. Therefore, the gun must never shoot wildly, but be always good alike. For ordinary sporting purposes a gun which shall give its largest killing circle at 30 yards with the first barrel, and at 40 yards with the second, will be found the most convenient for good shots.

Excellent shots may have the range extended ten yards, if possible, and indifferent shots the range reduced ten yards. It is the *ability* of the shooter to aim the gun which will determine the amount of choke. The fullest choke-bore will not "riddle" a partridge at twenty yards, but as the spread of the shot at that distance will only be half that of the true killing circle of the gun, a true aim must be taken. It is easier to correctly align a gun at twenty than at forty yards, and the indifferent shot, whose poor shooting arises from inability to aim, should avoid long shots. The poor shot who misses from nervousness should try long shots. A very little practice will give much more knowledge on the subject of a suitable choke than any number of instructions.

The shooting of a cylinder gun is so very poor that if the gun be intended for use with shot it should have *both* barrels more or less choked. The average pattern of a 12-bore cylinder is 120, and the idea of 120 pellets covering a thirty-six inch circle at 40 yards is too ludicrous. The distance between the first and last pellets is greater than in the choke, whilst at the regulation distance the pattern is pretty equally distributed over a four-foot target.

The writers of the Badminton Library advocate the use of the cylinder gun, and emphasise the word *good*, "as no guns vary more than do cylinders." Now, a barrel is a cylinder or it is not a cylinder—there is no question of degree; and the "Badminton cylinder," which makes "an average of 130 pellets with evenness, regularity, and penetration," is not a cylinder at all; it may be something in the guise of a cylinder. A cylinder,

12-gauge 30-inch, with standard charge, will make an average of 120. "Badminton" says that "choke-bored guns generally are rapidly becoming unpopular." The Proof House returns prove the contrary to be the case. There are some makers who cannot "choke" a gun properly; they are the "certain makers who, from the very first, have pointed out the doubtful advantages of full-choked guns for general shooting as applied to use in the field." The sportsman who requires a good shooting weapon will not patronise these makers, for it has been proved again and again that the choke-bore gun shoots *closer*, *harder*, and more *uniformly* than does the cylinder. It is a common fallacy that the cylinder is the best game gun; at the force gauge, the "plate," the "pad," or the "barn door," it is nowhere with the choke, but it is superior in the field. The old musket, costing ten shillings, and which only "goes off" once out of three times, has nothing to recommend it when compared with the modern breech-loader, save that "it is a good one to kill."

The author is willing to submit one of his full-choke guns to any test which may be applied to it, if the owner of any cylinder gun thinks he can devise any test which shall show the cylinder to be superior.

No sportsman will purchase a good gun without witnessing its performance at a plate, and he will reject a gun that with the same charge and same powder does not make a uniform pattern. If three close patterns are succeeded by one containing only half the number of pellets, the gun is most probably badly bored; but if the gun suits in every other respect it will be worth while to continue the trial, as it may be that a faulty cartridge has occasioned the poor pattern. If it be the fault of the gun, it will quickly recur.

It is sometimes advanced by the ignorant that the choke will wear; therefore buy a cylinder at once, as your choke may wear to one! It is fallacious to suppose that the choke, if rightly shaped and a true choke—that is to say, is formed by boring the barrel, and not by recessing, contracting, swaging, or other methods—will wear out. The choke-bored gun will, more probably, improve in its shooting powers. In 1875 the author's guns were submitted to a "wear and tear" trial by the manager of the *Field* Gun Trial, which extended over seven weeks, during which 2,500 shots were fired without any deterioration.

The author made a gun for Dr. W. F. Carver to use in his exhibition shooting, and upon trial of this gun after some 80,000 shots were fired from it the shooting was found to have improved. After 50,000 more shots had been fired the gun showed no signs of wear at the choke, nor had its shooting deteriorated.

In 1878 the author made a gun for the Schultze Gunpowder Company for their factory tests. This gun has had more than 80,000 shots fired from it; and as records of its performance have been kept, the average of the choke-barrel for each year can be given. It is quite certain that no Belgian barrel could have kept its shooting unimpaired so long.

Those gunmakers who habitually use Belgian barrels denounce choke bores.

Average of 1,000 shots per year :	Left barrel.	Right barrel.
1878	175	184
1879	202	196
1880	208	206
1881	207	201
1882	210	195
1883	214	194
1884	224	196
1885	239	201

These figures clearly prove that a choke will stand all fair wear and tear, and further that Schultze powder has no deleterious effect upon good gun-barrels.

The choke-bore gun has been particularly successful in the hands of both field and trap shots, and even its most bigoted opponents grant that in the hands of a good shot its execution is wonderful.

The following accounts of experiences with choke-bores will show the range and utility of a good choke-bored gun. The few instances cited must not be taken as extraordinary, but to quote many such in proof that they are not rarities would be a gross imposition on the reader :—

"On the 23rd of October, 1886, I was after ducks with my 12-bore, $7\frac{1}{2}$lbs. Greener Hammerless, and made three shots worthy of remark. With the first shot, I brought down two black ducks out of four, flying past (cross shot), at fully 75 yards ; the second shot, I killed a duck the size of a widgeon at 100 yards ; and with the third I knocked over a black duck at 110 yards. The two last were single birds."

The shot used in this case was chilled shot No. 2, there being only 138 pellets in the charge. The barrels of this gun are but 28 inches long.

"The first shot I had was with the 29-inch barrels, and I killed four pelicans in four shots ; not bad, considering two birds were all I could carry. . . I like the 31-inch barrels best, I killed with them a splendid turkey, putting nine No. 1 shot in his head, three in his neck, and five in his body."

This gun is also a moderately light 12-bore, by the author.

"With your gun loaded with $1\frac{1}{4}$ oz. of No 5 shot no single duck is safe at 80 yards. I have killed single teal at 80 and 90 yards, and on one occasion killed three times consecutively between 80 and 90 yards."

"With a full-choke Greener 12-gauge 8lbs. I shot at three wild geese 76 yards off, and I killed them all, using No. 4 chilled shot. A 'native companion' flew past. I killed it stone dead at 70 yards, putting no less than eight shots in its head and neck. This bird weighed $17\frac{1}{2}$lbs. On the same trip I shot at three single black swans at from 90 to 100 yards, and in three successive shots got them all. I killed a hare at over 90 yards with No. 4 shot."

"Mr. Cholmondeley-Pennell, who made the highest score, shot wonderfully well, at 30 yards from the trap, he killed all his birds with the first barrel. He used a full-choke by Greener."

"Last fall I killed two Brent geese, one with each barrel, distant over 100 yards.'

"With your gun I killed three wood pigeons, at over 70 yards' distance."

And so might ever-varying instances of the great killing powers of the choke-bore be cited. With large shot, its execution is still more marvellous. Particulars of shooting with buck and large shot will be found in the section dealing with the loading of cartridges, where instructions are given as to the best method of obtaining the *full* killing power of any particular gun.

ON THE FIT OF GUNS.

The fit of the gun is a truly personal matter. The sportsman must determine whether he will shoot with both eyes open, or one only; if the former, he must ascertain which eye is the stronger, and aligns the gun. If the gun is fired from the right shoulder and aligned with the left eye the aim will be faulty, and bad shooting naturally result. Happily, the great majority of persons naturally aim their weapon with the right eye, hence, all shooting correctors, cross stocks, &c., are useless to them. If there is any doubt as to which eye directs the aim, it may be easily ascertained by proceeding as follows :—

Take a finger ring and hold it out at arm's length; look through it *with both eyes open* at some object twenty or more feet distant; close the left eye. If the right eye still sees the object through the ring—which has not been moved—the right eye will align the gun, and the sportsman may with every advantage dispense with all correcting impedimenta, and shoot with both eyes open. If the left eye—being the stronger—aligns the gun, the sportsman must shut it, or shoot from the left shoulder; or have a particularly constructed stock which shall enable him to aim with the left eye whilst shooting from the right shoulder.

Providing the sportsman be one of the minority, he should write fully to an experienced gunmaker or the nearest practical gun dealer, and arrange for the building of a special gun, to meet his special need. If the sportsman be one of the majority, he may choose a gun that will fit by the following method :—

Take a gun and put it up to the shoulder two or three times without aiming at anything in particular; if it seems to come up easily, and to be under perfect control, choose a mark ten or fifteen feet distant, and slightly higher than the aimer's shoulder. Bring up the gun quickly, whilst looking at the mark, then glance at once down the rib, and see if the sight covers the object aimed at. Repeat this two or three times, and if each time the gun be found to be exactly upon the mark, try it at marks at different elevations; if it still covers the objects aimed at, it is a "fit," and a little practice will make the shooter feel quite at home with the weapon. To obtain a gun that will suit as well as that one, the gun may be sent to the gunmaker as a model, or the measurement must be taken according to the instructions given on page 44.

The object of the cast-off is to bring the mean centre of the barrels in a line with the shooter's eye, without having to press the cheek hard against the stock.

Too much cast-off is a great mistake. The same object can be gained by having a more crooked stock. The balance is an important point to be remembered in ordering one gun to be built like another. The gun should be balanced with a piece of thin twine, and the distance measured to the string from the breech ends of the barrels.

The sportsman who orders after inspecting a large stock of weapons has an advantage over the sportsman who, at a great distance from any gun depôt, has only a gun which does not suit him, and who wishes to order one that will. Sportsmen so fixed must read attentively the following remarks, and see in which particulars their gun fails.

In choosing a gun from the stock of any good maker or practised gun-seller, the sportsman will find many advantages. He will have the advice and assistance of an experienced man to guide him in fitting himself with a suitable weapon, and he will find an assortment of good weapons from which he may safely choose.

In taking up an ordinary good gun he will notice at once that the weight lies between the two hands; that it balances well; if tested it will be found to balance within 3 inches of the breech. It is easy to make a gun balance by loading the butt, but this clumsy expedient does not make a well-balanced weapon of it, for the weight being at both extremities, instead of between the hands, it will be found to answer better as a balancing-pole for immature Blondins than as a sportsman's weapon.

The good gun when placed on end will be found to stand upright, or nearly so. The shape of the heel has much to do with the fit of the gun; too much heel prevents the gun being brought readily to the shoulder, and causes it to shoot *lower* than a gun with a full toe.

The butt-plate will not be very much rounded, and will slope in a little to fit the shoulder—that is to say, measured to the *edge* of the heel-plate, the stock will be very slightly longer on the off than on the near side. The heel-plate will be roughed to prevent it moving when held against the shoulder. There will be room between the two triggers for the thickest finger, and both triggers will be so placed as to be readily reached by the trigger finger; the hand will be tapered from the head to the extremity of the grip, to prevent injury to the second finger by the guard as the gun recoils. Bruises of this kind are most unpleasant, and far too frequent.

The fore-end will be high, narrow, and about seven and a half inches in length. The grip small.

In many gun-shops will be found a dummy gun, which may be adjusted to any length, bend or cast off, and which greatly aids in quickly fitting a person with a gun. The sportsman may not, although a first-rate shot, hold or bring up the gun to the best advantage. The gun should be grasped with the left hand as well forward as convenient; and in trap shooting it is held well up and the butt brought up to the requisite elevation.

In ordinary shooting it must also be got well forward, and the gun thrown slightly from the shooter and brought against the shooter; immediately it touches the aim should be correct, and the trigger pulled.

A long stock is held more firmly to the shoulder than a short one, and the shock of the recoil is thereby lessened; if too long, the gun requires more time to be brought into position, and shots may be lost on this account.

The pistol grip enables a firmer and more comfortable grasp being taken, and usually the gun is more tightly pressed against the shoulder with this pattern stock.

Americans and English colonists generally make use of very crooked stocks; the English and Continental sportsman prefer straight stocks.

If used to shooting with the body and gun at right angles, and without bending the neck, a well-bent stock will be found advantageous. There is no necessity for Englishmen to use such straight stocks as they do, but so long as they shoot in the usual English style, with the shoulder well forward, and the neck bent, a straight stock will seem more convenient.

All these points in the measure of the gun must have attention if the sportsman is to be suited as he should be. It is apparent that any machine-made gun will never allow of the weapon being adapted to the shooter. The sportsman must accommodate himself to a poor specimen of the sporting gun, an ill-balanced thing too often, and always inferior to a best hand-made gun. From the use of it he will not probably derive much comfort. The idea of a machine-made sporting gun is crude; it is as if all the world could make comfortable use of *sabots* specially designed for Australian aborigines.

The following short *résumé* of the points to be looked for in a gun will be of service :—

The gun must be of weight and calibre proportionate to the strength of the user. The amount of choke will depend upon the purpose for which the weapon is intended and the skill of the user. The gun will be central-fire, with a simple mechanism of breech-loading, the lever being upon the top.

If hammerless, the gun will be more expensive, and a perfectly simple lock and cocking mechanism will be demanded.

For pigeon shooting, for wild fowling, for all light guns, for all guns with extra long chambers, the Greener cross-bolt will be made a *sine quâ non.* The barrels will be of best twist, or "sterling," steel, the figure regular, small, and light in colour; the breech-ends heavy and stout to the end of the fore-end; they will be polished as a mirror, inside and out, and will have a transparent, rich colour upon their exterior. They will bear the English proof-marks. The length of the chamber will be that usually employed for cases *readily* obtainable.

The breech-action will work smoothly, and require no extra pressure at any point of its travel. The locks will be front-action, and so constructed as to permit of hammers being below the line of sight when at cock, and the mainspring so shaped as to cut but little into the breech-action body. The fences will be elegantly shaped, and a true pair. The strikers will be without springs.

The pull-off will be light and free. The hammers will be small, elegant, and have a long travel; the triggers broad and widely spaced; the stock in dry wood, straight-grained at the grip, curly and deeply veined on the butt; the comb round; the butt-plate broad and deep and roughened; the fore-end easily detachable.

The following will be absolutely refused:—Guns of intricate mechanism, cheap light guns, cheap hammerless guns, all guns of which a trial is refused, all guns not bearing a proof mark, all guns marked "London fine twist," or named with a dead or fictitious name, all guns in which the barrels slide forward, or turn over, or move on a vertical axis for loading, all guns worked by means of a lever under the fore-end, pin-fire guns, guns having sham barrels, hammerless guns with more than two safeties, guns having barrels brazed from end to end, or soldered together with sal ammoniac, guns with crooked barrels, dented barrels, crooked ribs, or ribs more on one than on two barrels, guns which never fire, or spit at the breech when fired, guns with stamped vulcanite heel-plates. The following will depend upon the fancy of the purchaser:—

The rib, whether flat or grooved, and whether plain or deadened by engine-turning, file cutting, or engraving.

The amount and style of the engraving.

Whether pistol hand, or straight hand, or half-pistol hand.

Whether without butt-plate, and the butt chequered; whether with tips-and-ends, or a horn chequered butt-plate, or a metal butt-plate.

Whether or not furnished with cheek-piece, swivels for sling, and escutcheon for monogram or crest.

COMBINED GUNS.

Shot guns can be fitted with rifle barrels, but, as a rifle, the weapon must be heavier than it would be if it were constructed for a rifle simply.

The guns may have extra barrels; but the barrels should be of the same bore, and without great difference in weight. Guns built as 10-bores, and fitted with extra 12-bore, barrels cannot be recommended, neither can such makeshifts as detachable choke-bored muzzles, and Morris or other tubes for converting shot guns into rifles.

SECTION VII.—PRACTICAL.

HOW TO USE THE GUN.

It will be perhaps as well to state at once that it is not the writer's intention in this book to give any hints as to the art of shooting. The use of the gun is to comprehend the handling and care of the weapon—nothing more.

The rudimentary directions generally given by all would-be instructors of young sportsmen are hardly necessary in this small work, but the author cannot but reiterate the caution—"Never point a gun at any one," and reiterate also the hope he has so often expressed that the pointing of a gun at any person shall in itself constitute a legal misdemeanour.

The gunmaker's motto, "Make all sure," is the best sportsmen can adopt. Have a care as to the direction of the muzzle—always. Carry the gun loaded, and at full cock, and if it be a hammerless gun, with the "safety" off. In getting over a fence, creeping through a gap, or jumping a dyke, put the hammerless gun to "safe," or *unload* the hammer gun. It is tiring to carry, even the lightest gun, long in one position. The method of carrying may therefore be varied—over the shoulder, across the breast, or muzzle down, as the nature of the ground or position of shooters may determine, but in changing the position of the gun take care that the muzzles do not cover any one.

In pigeon shooting the rules enforce the loading of the gun when the shooter is at the mark. Ordinarily the gun is taken from the rack or given to the shooter by an assistant. As soon as fired the empty shells are withdrawn, and the gun returned to the rack. A safety is of no use whatever on a gun to be used solely for pigeon shooting, and its presence may lead to great annoyance.

The shore shooter may often exercise greater liberty as to the manner in which he shall carry his gun on his solitary rambles, than can be permitted in the field; but he must never be careless. In shooting from a sailing boat the gun, when loaded, should never be without the sportsman's hand upon it.

When that method of shooting usually called "walking in line" is practised, the sportsman must carry his gun with the muzzle up. In partridge driving the sportsman has often a pair of guns, one of which will be always in his hands, the other in that of the loader. The adjoined illustration, which is reproduced from *Shooting*, by kind permission of the publishers of the Badminton Library, shows the position of loader and firer, and further exhibits to the sportsman a gun in the *correct* position for firing.

When a loader is employed he should be instructed to load the gun with the muzzle directed to the ground, and he must bring up the stock of the gun to the barrels when the cartridges are in, so that if by any accident the gun should go off the shot will bury itself harmlessly.

Shooter and Loader.

More breech-loaders get shaky in the action by being worked carelessly than from repeated firing or the use of heavy charges. The barrels of a breech-loader should never be jerked down, nor should they be thrown back into position with a snap. The proper manner in which to load a gun is to drop the stock under the elbow, and press it firmly against the hip or the body, unfasten the lever with the right hand, and with the left grasping the barrels a few inches in front of the fore-end, lower their easily. Close the gun in a careful manner after fitting in the cartridges, bringing the stock up to the barrels.

Guns must be taken care of; they will then last many seasons, and give every satisfaction. When not in use they should be stored in a dry cup-board, either on a rack or in their travelling cases. The gun must not be put away at "full cock;" the tumblers may be lowered by pulling the trigger, and such is better than allowing them to remain at full cock; but if the gun is laid by whilst together, two used cartridge cases, well greased, may be put in the chambers, and the locks snapped off; or the tumblers may be lowered by pulling the triggers whilst the barrels are open, but this cannot be done with guns fitted with automatic trigger-holding safeties, nor with hammerless guns which are not cocked by the dropping of the barrels. The gun before being put together or used, should be wiped free from dust; nothing tends more to clog the breech mechanism. Having put the gun together, it should be opened and shut several times, and any stiffness or clogginess will at once be noticed. If the gun is a snap action the lever should be carried quite home as the gun is closed. Dirt often finds its way *underneath the extractor*, and this even in a most minute quantity will frequently occasion stiffness in working, or very possibly prevent the gun from closing. Oil and dust, and sometimes a little rust, will be found in the bottom holding-down bolt; this causes the gun to work stiffly. The gun must *never* be *forced* open, or unusual *force* used to close it. If the gun does not open *freely*, it should be carefully examined, and on the principle that a stitch in time saves nine, it may be cleaned thoroughly, providing the cause of the stiffness is not found, and the obstruction removed. In putting a gun together, providing all the parts are clean, no stiffness will be noticed and no force requisite. In case of a *deadlock* in putting in the barrel, do not attempt to force the barrels in, but search for the *cause*. Probably, if a hammerless gun, it will require cocking; if a hammer gun, possibly the strikers are projecting through the face, and do not work freely, so that the extractor drops upon them and prevents the barrels going home. The keeping of the gun clean, and the mechanism free from grit, will ensure immunity from the annoyance of a "jam" in the field. In case of the sticking together of parts that should work freely—such as the strikers jamming in the breech-action, the extractor clogging in the barrels, or bolts or any parts becoming fast with rust—there is nothing so good as an application of petroleum; repeated applications, and the exercise of patience, will not fail to loosen the "cement," and make even the rustiest pin amenable to the persuasion of a hand turnscrew. Having the gun together, and working freely, it will require to be used carefully. It must not be let fall heavily on its butt plate; it must not be pushed underneath the seat of the dog-cart or waggonette, and left to take its chance; it should not be left muzzle-up or muzzle-down against a wall, a gate, or a tree. It should not be used as a crutch, an alpenstock, or crowbar. From a critical examination of many guns returned to the author, after very little wear, he fancies they must at times be utilised for very different purposes to those for which their makers intended them. To speak more plainly, some guns are abominably abused. Most often those men who make such ill-use of their guns are the loudest in their complaints if any part gives way. A gun of good quality is a finely constructed piece of mechanism; the strain it has to withstand at each discharge has been ascertained to average 40,000 lbs. Many guns are fired from ten to twelve thousand times in one season, and will last, with ordinary care, fifteen or twenty seasons. The guns which in reality have the least wear—pigeon guns—are usually cared for best. Certainly the pigeon gun has to stand heavier charges than are employed for game shooting, but the number of shots fired will bear no comparison to that used by an enthusiastic English sportsman in one season's game shooting.

The more expensive the weapon the more perfectly fitting are the parts. Many shooters seem to think that because a gun costs £30 it will withstand any amount of ill-usage. The contrary is the case; a badly fitted, heavy, cheap gun, will make a better crowbar, but it will certainly break down before it has been fired a thousand times, and may burst at the *first shot.*

Without advocating the laying-by of guns, or treasuring them as works of art "too good to be used," they may be cared for intelligently: kept clean and dry, the working parts lubricated occasionally, and shot as often as possible.

In most hammerless guns, if the *cover plate* underneath the breech action *body* is taken off, the locks may be inspected, oiled, and any rust, or clogged oil and dust, removed from the *bent.*

If practicable, have a good gun examined each summer by its maker or a practical gunsmith.

Never send a good gun to an advertising jobster, who is prone to stick his name upon your gun, under the pretence he has improved its shooting.

To have a gun "choke-bored," or to have it converted into a cylinder, always send to the maker or a first-class gunsmith, and do not be persuaded to let any jobbing man tamper with your gun-barrels under any pretence.

In sending a gun for repairs send the *whole* gun. If shooting requires to be regulated, choke-boring removed, chambers altered, it is absolutely necessary that the workman has the whole of the gun if justice is to be done to the work, and for the trifle additional in cost of freight it is always best to send the gun complete.

Other sportsmen like to take their guns all to pieces and re-arrange the parts. This is not requisite, and does not in any way add to the efficiency of the arm. The gunmaker is the proper person to take apart the locks, or strip the breech-action; if there is not a practical man within easy reach the sportsman must, of course, himself endeavour to effect any repairs, but it is not advisable to interfere with any gun that functions properly, nor to practise upon any gun that works satisfactorily.

To clean a gun after a day's shooting. If the gun be wet, it should be wiped dry *at once*, but the *cleaning* of the barrels and breech-action may be left until the sportsman or his servant has time to do it properly.

To clean the barrels. Use the cleaning-rod, with tow and oil, or turpentine. To remove the fouling, put muzzles on the floor, and push the rod down to within an inch of the muzzle, and draw up to the chamber. Do this two or three times, then push right through. Use the bristle brush, or the rod with plenty of flannel; finish with the mop soaked in refined neatsfoot, pure Arctic sperm oil, or vaseline.

Never half-clean the barrels; always wipe them dry and clean before finally oiling, and do not put the mop used for oiling into a foul barrel.

Always wipe the *bed, face, and joint* of the breech-action with an oily rag or flannel. A little linseed oil may be rubbed over the stock occasionally.

Before putting the gun together ascertain that all the bearing parts are free from dust or grit.

The *joint* may be lubricated with a mixture of half best Russian tallow and half petroleum.

The cocking-lifters of hammerless guns, the holding-down and top bolts, and the triggers, if they have a tendency to clog, may be touched with a feather dipped in petroleum. They must be lubricated, whenever they require it, with chronometer oil, Rangoon oil, or finest neatsfoot.

Rusty or tight breeches in muzzle-loading barrels may often be turned out, providing the breech-ends of the barrels have been soaked in petroleum. Very obstinate breeches may require to be well heated, as well as lubricated, before they can be turned out, but usually petroleum will be found a sufficient remedy for incipient rust of the working parts. All the parts of the mechanism may be cleaned with petroleum; it removes clogged vegetable and animal oils well. Turpentine, often used successfully to clean the residue from gun-barrels, will give great trouble if it get into the fine-fitting parts of the mechanism of the breech-action and locks, and must therefore be used with care.

To remove leading from barrels, roll a little quicksilver up and down the barrel.

To remove rust from inside or outside of a barrel, procure a tub, and with a kettle of boiling water well scald the barrels inside and out, wipe perfectly dry with flannel, and then oil. It is as well to do this before putting the gun aside for any length of time.

When laying a gun by, or packing it for a long journey by sea, after having well dried the barrels, oil inside and out with a heavy-bodied oil or melted tallow, free from salt or acid: stop the barrels with oiled corks or wads, or put in baize-covered sticks. Smear the breech-action and metal furniture with grease.

STRIPPING AND REPAIRING GUNS.

To take to pieces a breech-loader for cleaning or repairs, first remove the fore-end and barrels; then, with a strong hand turnscrew, turn out the side-pins, and remove the locks and hammers together; next turn out the guard-pins, and remove the bow or guard; another pin will then be seen in the rear end of the trigger-plate; remove this pin (occasionally this "hand-pin" is placed in the reverse way, the head of this pin will then be found on the top of the grip in the tang of a long break-off). The "furniture-pin" should next be partly turned out; this pin fastens the fore part of the trigger-plate to the body of the breech-action, and is easily distinguished. Next remove the "breech-pin" upon the top of the tang of the break-off; in top-lever action guns the breech-pin is covered by the lever, which must be held on one side whilst the pin is being turned out. Rarely a false pin is screwed into the lever, which, when removed, will leave an aperture through which the breech-pin must be extracted. After having removed the furniture-pins, the trigger-plate and triggers may be taken from the stock, after which the breech-action may be removed entire.

To strip breech-actions, if the action is a treble wedge-fast or ordinary top-lever double-bolt action, the first thing will be to remove the spring. To do this, first partly turn out the lever spring pin (under tang of break-off), and with a pair of pliers or pincers take hold of the spring and slightly grip it, and lift the spring towards the head of the pin. It will then be free from its bearing, and may be removed by completely turning out the spring pin. (This does not apply to spiral springs.) Next proceed to turn out the pin or pins connecting the tumbler with the bottom bolt, and remove the bolt by drawing straight out backwards. Next turn out the lever pin on top of lever, and by means of a small wire punch inserted in the lever pin-hole, knock out the tumbler. The lever may then be removed, and the top bolt, if any, will fall out. In side-lever guns first knock out the pivot on which the lever works, then remove spring and bolt. Snap guns with lever under-guard may be stripped in much the same manner, but the spring and lever are fixed to the trigger-plate, and the spring must be removed before knocking out the pivot-pin. Owing to the numerous complicated breech-actions that are made, we are unable to say that the above directions will be sufficient to enable an amateur to strip his gun; but we trust they will be explicit enough as regarding guns of our own make. There are many breech-actions made that puzzle expert gun-smiths to take apart and repair, and it would be foolish for an amateur to attempt to take them apart if a gunmaker is within reasonable distance.

To strip a muzzle-loader, first remove the lock, then the barrels, then proceed to remove the furniture and break-off, as already described for breech-loaders. In military rifles, the bands fastening the barrel to the stock must be loosened by a screw underneath, and then removed by slipping over muzzle of barrel. (Note.—Horn heel-plates are usually glued to the stock, as well as being fastened by the screws.)

To strip a gun-lock, first remove the mainspring. This may be accomplished with a pair of lock vices, or a cramp may be made by filing a notch or slot in a narrow strip of $\frac{3}{16}$ iron or steel, the size of the breadth of mainspring when at full cock. Having cocked the lock, slip the cramp up the mainspring until it catches, then release the scear and push down the tumbler. The spring being firmly held in the cramp, it may be unhooked from the swivel and removed from the lock-plate; then unscrew the bridle-pins and remove the bridle.

The scear may then be lifted off if the tumbler is not in bent. The scear spring will then be at liberty, and may be removed by turning out the pin. Now the hammer should be removed; the tumbler-pin is first turned out, and by means of a wire punch inserted in the hole, the tumbler is knocked away from both hammers and lock-plate. If a hammer fits well, it will be impossible to remove it in any other way without injury either to the hammer or the lock. The spring must not be taken out of the cramp; it requires no cleaning except at the claw or hook. In putting a lock together, first screw on the scear spring, then the tumbler, then place on the scears and cramp the spring with a pair of pliers or tongs, place the tumbler into half-bent. Then affix the bridle, and screw it to the lock-plate. Take the mainspring, ready cramped, hook on to the swivel in tumbler, place the stud in the hole drilled for it in the lock-plate, raise the tumbler to full bent, squeeze the mainspring down close to the plate, and remove the cramp; the lock will be ready then for affixing the hammer, which should be knocked on after placing the lock firmly on a solid block to prevent the bridle from breaking. To take apart the lock-work of the Anson and Deeley Hammerless Gun, proceed as follows:—

Having removed the barrels, snap down the hammers or tumblers, remove the cover-plate from bottom of breech-action body; knock out with a wire punch, from the *right* side, the scear pivot, or the one nearest the stock, and remove the scears; knock out the dog-pin, or the one nearest the fore-end joint, and remove the cocking levers; partly screw on the cover-plate, and carefully knock out the centre-pivot or tumbler-pin, remove the cover-plate, and the tumblers and mainsprings will drop out upon the breech-action being reversed. The scear springs lie along the bottom of action, and may be removed after turning out the pins. To put the lock-work together, first place the mainspring in the bend of the tumbler, with the stud of mainspring bearing in its proper slot, and its other extremity bearing against the under side of the nose of the tumbler; the tumbler and spring having been placed in the slot must be forced into position with a cramp, or piece of notched wood; knock in the tumbler-pivot half-way, insert the other tumbler and spring in the same manner, and knock the wire pivot right through the lifting dogs, and scears must then be put in, and the whole covered with the cover-plate. The "Greener" hammerless guns, which have similar tumblers and scears and mainsprings, may be taken to pieces in the same way, but there are no dogs or lifters to be removed. Illustrations and descriptions of all the parts of the locks mentioned here will be found on page 38.

In putting guns together, proceed in the reverse order to that described for taking apart.

The best methods for temporarily repairing guns when hunting beyond the limits of civilisation will now be detailed; but in cases where possible, it is advisable to send the gun to a practical gunsmith for all repairs. A list of foreign and colonial dealers is appended, and the repairs done by them may be relied upon. The following extra parts of a gun should always be taken when out hunting, or exploring in wild countries :—Extra pair of lock mainsprings, extra hammers or tumblers, pair of extra scears, pair of extra nipples and strikers, extra side-pin, extra action or lever-springs. It is possible that the lever-spring may break, but it will not in any way affect the utility or safety of the arm, only the lever will have to be moved home when the gun is closed, instead of it snapping there. The strikers of ordinary guns will become useless after continued wear, owing to the hardened hammer flattening the head of the striker, and so shortening its travel as to make miss-fires of frequent occurrence. The nipple must then be turned out with a key or a pair of pliers, and a new spare striker inserted. In hammerless guns the tumbler and striker being in one, and the point itself striking against the soft copper-cap of the cartridge, this flattening does not occur, the strikers being of the best mild steel carefully hardened and tempered, and so well made that breakages are of very rare occurrence. We have had one gun fired over 10,000 shots without the tumbler being any the worse for wear or rust. Perhaps the most usual accident to a sportsman will be the denting, or the bulging of the barrels. When a bruise is discovered, DO NOT IN ANY CASE SHOOT out of the gun until the barrels have been repaired, if the bruise is a bad one ; for firing out of a badly bruised barrel invariably causes the barrel to bulge considerably at the bruised part. To remove a dent, the following is the readiest expedient :—Having removed the barrel from the action or stock, insert in the barrel at the breech-end a solid leaden plug or bullet, as near the size of the barrel as possible, place the barrels on a solid block with a stout ramrod or stick in the barrel, reaching within a few inches of the chamber, then proceed to flatten out the plug or bullet by striking it with another rod and a hammer ; the bullet being prevented from slipping down the barrel by the ramrod underneath, it will expand until it perfectly fits the barrel ; then proceed to force the plug—having first lubricated it—towards, and gradually past the bruise, turn the plug half round in the barrel, and repeat the process until the bruise is raised. The barrel should be warmed during the process by applying a hot iron to the outside of the bruised part. Great care will have to be taken not to get the plug jammed in the barrel. If a taper lead plug can be obtained, the process will be greatly simplified, and a slightly taper iron or brass plug is much better than a soft lead one. If the barrel is bulged a similar plug should be made, and great care will have to be taken to hammer the bruise down to the plug with a light hammer. If a hard metal plug can be obtained near the required size, it may be packed with paper until of the required diameter. The plug must be slightly longer than the bruise or dent. Another frequent accident in wild countries is the breaking of the gun-stock. This may be securely spliced in the following manner :—First glue the stock as well as possible, then glue round the fracture a piece of thin leather or canvas, and whilst warm tightly bind with waxed thread or a fine lace; when the whole is dry it will be almost as sound as before. The wood should be warmed before gluing, to enhance the chances of perfect success.

Repairs to breech-actions require great care and experience in effecting, and always when practicable the gun should be sent to the maker, as he has more interest in properly repairing it than any one else. To tighten a breech-action, the usual way is to fit a new hinge-pin slightly larger than the old one, or by filing from the flats beneath the barrels, and hammering up the bites on the lump, which process brings the breech-ends of barrels nearer to the face of the standing-breech. When the cartridge bursts at the rim at the upper edge of the case, it is a sure sign that the gun requires to be tightened up.

ALTERING GUNS.

Guns cannot be converted from hammer to hammerless upon any good reliable system; they cannot be made self-ejecting on the Needham or Greener principle, and with the exception of converting non-rebounding locks to the rebounding principle, it is doubtful if any conversions pay for the cost.

CARTRIDGE LOADING.

It may be of advantage to repeat here the information given more fully in the various technical portions of this treatise respecting the loads with which guns may be charged to best advantage. The standard charges for the guns of different gauges will be found detailed on page 57, and provided that No. 4 Pigou, Wilks, and Laurence's powder is used, these loads cannot be beaten for general purposes. These loads may, however, be varied for certain purposes, and a reference to the fac-similes of Killing Circles in Section V. will show the purpose of certain variations.

Sportsmen must, however, remember that for closeness of shooting the Schultze gunpowder is without an equal in guns of ordinary calibre, and in which paper cases are used. In guns of 8 and 4-gauge, as also in 10-bores, when used with brass cases the superiority is not so marked. In brass case guns—that is, guns bored and chambered especially for Kynoch's thin brass cases—which guns have little or no chamber cone, no nitro-compound will shoot closer than black gunpowder, and for use in such guns, as well as in cylinder guns, unless the smokeless quality of the nitro-compound is deemed an advantage, it would be better to use black gunpowder of No. 4 grain. The size of shot to be used must to some degree be regulated by the bore of the gun, as well as by the size of the game at which the charge is to be fired. The 28-bore gun will do relatively better with 8 or 7 than with 6 or 5, and with a 4-bore gun anything smaller than No. 3 is wasted.

Modified choke-bores and cylinders give a larger killing circle the smaller the shot; 28-bores will kill ordinary winged game at 40 yards with No. 8, and 20, 16, and 12-bore cylinder barrels may be used at the ordinary winged game of Great Britain with greater success if charged with No. 7 than with 5 or 6. In 12-bore guns it is not wise to load with larger shot than No. 4, unless the gun has been regulated for use with a certain size of large shot. For 10-bores No. 1 size is the limit. These remarks must not be construed to mean that the 12-bore game gun must always be used with one size of shot; for quail and young partridge shooting No. 7 may be employed to advantage, and for shore shooting even larger sizes than No. 4—but the sportsman must remember that when he uses the large shot he is sacrificing closeness of shooting for the extra benefit he obtains from the increased range and smashing power of the large shot. Cartridges loaded with nitro-compounds must be well turned down.

Good close shooting in guns of any bore can only be obtained by using cartridges loaded rationally, and to be rationally loaded there must be good wadding between the powder and the shot. The secret of good shooting is in the employment of a first-class felt wad over the powder; and it is imperative that this wad be of good quality. The texture must be close and firm, but the relative hardness or softness of the wad is of less moment. It should be of the same diameter as the internal diameter of the cartridge case in which it is to be used. To protect the *powder* from the injurious effect which may result from continuous contact with the chemically prepared felt wad, a thin card wad or a waterproof wad should first be inserted; and it is supposed to be conducive to closer shooting if this protective wad, instead of being a simple card wad, be a compound paper and felt wad—technically known as the "pink-edged" or "FIELD" wad. This "Field" wad should always be used when loading with nitro-compounds; with black powders its use is not so imperative. It is customary to place a third wad, of thin card, between the powder and shot, but it is very doubtful if any benefit will accrue from its use.

The cartridge loaded for close shooting will therefore be charged as in this illustration.

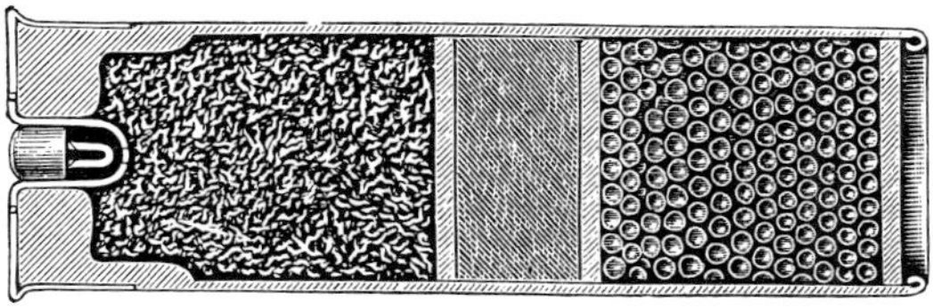

Cartridge loaded for Close Shooting with Black Powder.

Loading with two pink-edge wads over the powder, and one pink edge wad over the shot, as is often done in the United States, causes the charge to scatter, and such loading will lower the pattern 10 per cent. in a gun fully choke-bored.

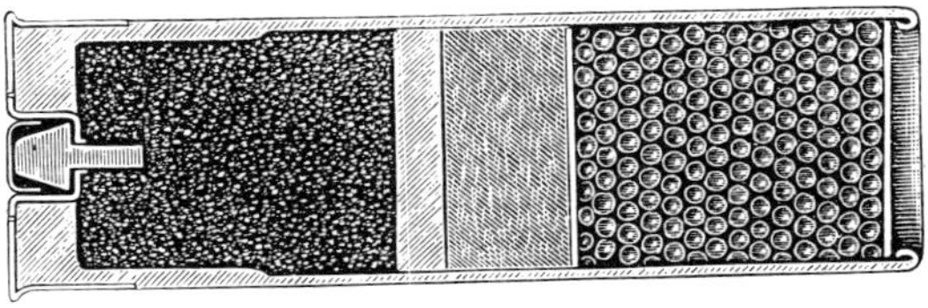

Cartridge loaded for Close Shooting with a Wood Gunpowder.

If charged with "Schultze" or "E.C." gunpowder, the wads used will be the pink-edged or "Field," the thick felt and the thin card as shown.

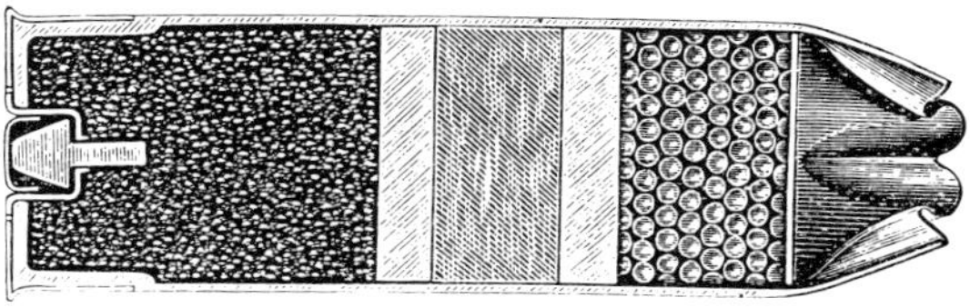

Brass Cartridge Case Wadded for Close Shooting.

If brass cases are to be loaded for close shooting, put the wads, as illustrated, between the powder and shot, and crimp the case.

To load choke-bore guns so as to scatter the shot at close quarters has

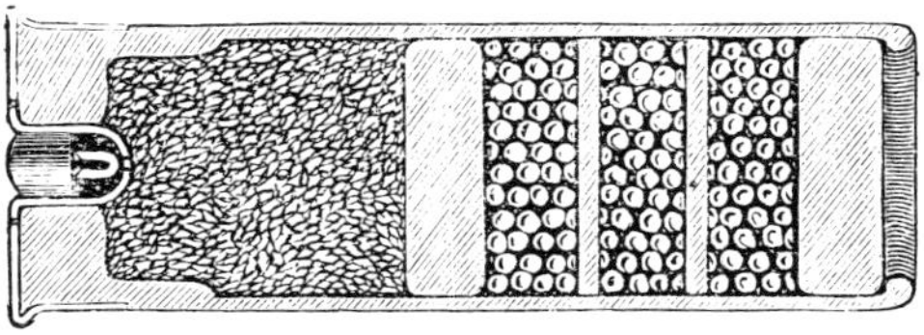

Cartridge loaded to scatter the Shot.

always been a difficult matter to accomplish. The rule is to diminish the thickness of wadding between the powder and shot, and increase it over the shot. This is pretty effective, but the best plan is to load as illustrated.

The charge of shot, it will be seen, is separated by two cardboard wads. This will cause a full-choke-bore gun to make a pattern of 140 at 40 yards instead of 220. A still smaller pattern may be obtained by using one ounce of shot instead of the ounce and eighth, and still further by substituting No. 5 for No. 6 shot. If it scatters too much, separate the shot by one wad instead of two, or by simply using one pink-edged wad only over the powder, and one or two over the shot.

The scatter-charge has good penetration at 30 or 40 yards, but of course not so much as when loaded for close shooting.

For rabbit shooting, with a full-choke-bore gun, at 15 and 20 yards' distance, very successful results have been gained by reducing the charge of shot to $\frac{1}{2}$ oz., the smashing of the game being thus avoided.

There is an idea prevalent that by diminishing the charge of powder below the standard charge, and keeping the standard charge of shot, the closeness of the pattern will be increased. This is erroneous, so far as it applies to guns choke-bored, and the use of black gunpowders, and in any case it is productive of irregular shooting and patchy patterns.

By over-loading the gun with powder and shot, or with powder only, irregular patchy and open patterns are produced.

When small charges are used the cartridges should not be shortened, by cutting down, or turning over beyond the usual limit, but by filling the space with wadding—or the case may be slightly turned down, and then crimped, as are the thin brass cases.

It will be found to be true economy to purchase just such cartridges as are required for the sport purposed than to use unsuitable loads. For partridge shooting use the *Sporting Life* cartridges; they are also suitable for grouse shooting, and may be looked upon as amongst the best for all round purposes. For shooting in pigeon matches and at wild fowl use the very best cartridges procurable, and by no means employ reloaded cases.

For rabbit shooting a cheap cartridge may be purchased, but such cartridges should not be stored. It is economical to work up old, damaged, mixed, or suspected cartridges at a rabbit shoot.

In Great Britain it is generally cheaper to purchase cartridges ready loaded than to attempt to load the cases, and it is certainly cheaper to purchase from a London or Birmingham gunmaker than to order from the local dealer. It will even pay to have small lots of even two hundred sent by rail hundreds of miles, although the cost of carriage on five hundred and one thousand lots is proportionately less, and if the shooting warrants the purchase of cartridges in five hundred lots they should be so ordered. The best cases should be ordered, unless the cartridges are to be shot off at once. The brown cases offer but little inducement as to price, and they must be used at once if satisfactory results are expected.

By purchasing in London or Birmingham the buyer ensures the lowest price if he is prepared to pay cash; he gets new goods, powder, wads, and shot of guaranteed quality, and the loading carefully and accurately done, and thus uniform results are ensured. The cartridges obtained locally are often old stock, generally loaded or partly loaded at one of the large cartridge case factories, the powder being of the ordinary trade quality, the shot wanting in evenness and rotundity, and the price high. There is also the probability of the load or size of shot required not being procurable at any price. Loaded cartridges in lots of two thousand or more can be sent economically to most of the English Colonies, to the United States, and to several foreign countries. The author will be pleased to give any information he may possess or obtain as to the shipment to any port of shot or rifle cartridges, should any person wishing to import goods of this class apply to him. Gunpowder is not so easily shipped as cartridges, but the quality of English gunpowder is so favourably known, and the best brands are in such demand, that the leading manufacturers have depôts at the chief ports of those countries into which the importation of gunpowder is permitted. Gunmakers who make a speciality of loaded cartridges ensure the most perfect accuracy in loading, and by the use of particular machinery and accessories can promise a degree of excellence neither obtainable by the ordinary method of hand loading, nor by the purely mechanical methods practised in cartridge case factories.

The sportsman may occasionally find it necessary to load his own cartridges, and will often do it in preference to trusting to the country ironmonger's shopboy; but he can never do it as economically, nor with such accuracy, as the gunmaker, who has every mechanical contrivance, and well-skilled workpeople to assist him. To load quickly and accurately, place the cases to be loaded base down upon the table. Adjust the powder measure, put the powder into a basin, take up a full measure, strike it off level with the base of the case to be loaded, and pour it in the case carefully, holding the case over the basin. Having filled all the cases with powder, put in the first wad and press it squarely and firmly down; the succeeding powder wads may then be put in and rammed home together. The shot must be poured in through a tun-dish, and preferably counted with the "Greener Shot Counter," or weighed to measure—the top wad of thinnest card, and not too light; the turning over firmly and evenly done. By this method it is possible for one person unaided to load well and accurately three thousand cartridges in a day of ten hours.

Cartridge cases do not pay to reload; it is false economy in England to reload paper cases or perfects—abroad it may be necessary to do so occasionally, but no case fires so well a second time. Paper and brass cases both quickly lose their elasticity if reloaded and fired time after time, and in reloaded cases there will always be a greater percentage of miss-fires than in new ones. The de-capping must be done as soon after firing as convenient. New anvils will be required to some cases, and care must be taken that the cap and the anvil are both got well "home" in the cap chamber when re-capping, or miss-fires will certainly ensue.

Paper cartridges and brass may be kept over from one season to another without appreciable deterioration, providing they are stored in a dry, airy place, and at an equable temperature of 50°. Powder, whether black, Schultze, or "E.C.," that has once become damp or sodden, can never be dried so as to recover its original strength. Some nitro-compounds may be over-dried and develop dangerous qualities, but with either the Schultze or

"E.C." powders it is not probable that such will result with even the most careless storing. It must not be supposed that the powder in damp cartridges can be dried without unloading the cartridges; to attempt to do so would be like trying to dry one's stockings without taking off one's wet boots.

SELECTING BUCKSHOT.

To select a buckshot which will suit a particular gun, put a wad in the muzzle of the gun, about half an inch down, and fit the shot in perfect layers. There will be no need to try them in the cartridge-case, as it is immaterial how they may fit there. Buckshot of a size which will average nine pellets to the $1\frac{1}{8}$ ounce will generally fit a full choke-bored 12-gauge gun, the three shot to the layer. If a smaller shot be required, choose four or five to the layer, avoiding sizes which do not fit fairly well. With such well-chosen buckshot the whole of the 9, 12, 16, or 20 shot of the charge will be in a centre of 25 inches diameter at 40 yards range.

THE GUN-ROOM.

Guns and shooting paraphernalia should be kept together. If a room cannot be devoted solely to them, a capacious cupboard, or a case fitted with a gun rack, and several drawers and shelves, will contain a small battery and the requisite accessories.

Guns are best kept put together and placed butt down on a gun rack in a glass case or gun cupboard, but if the case is not practically dust-proof, the guns should be first put in pliable canvas or cloth covers. Guns kept in racks in the open room should always be kept so covered.

Loaded cartridges are best kept on an open shelf, and in a current of air; boxed up in an air-tight cupboard, they will deteriorate more quickly.

After the close of the season, inspect the guns very closely, and send those concerning which there is any doubt to the gunmaker for repairs at once.

On receiving his report it will be as well to decide quickly whether or not new weapons must be purchased for the next season. Some wet summer day overhaul the contents of the gun-room, put the odd cartridges handy for popping at rabbits or vermin, see that the cleaning tools are complete, that the cartridge bags, game bags, &c. &c., are in good condition, and make a list of the things which will be required when the season opens.

In the season the gun-room will require frequent attention if it is made use of by more than one person. The cartridges, as soon as they arrive from the gunmaker's, should be transferred to the magazine or cartridge bags of the shooter for whom they are intended; a cleaning-rod and gear, turnscrews and extractor put in the travelling gun-case, and the oil bottle refilled.

PART II.—AMMUNITION AND ACCESSORIES.

SECTION I.—GUNPOWDERS.

THE explosive used in shot guns is either black gunpowder (saltpetre, charcoal, sulphur) or a nitro-compound (carbon base, treated with nitric and sulphuric acids). The black gunpowder is granulated, the grains being of various sizes, ranging from dust to 1½-inch cubes for use in cannon.

The ingredients of best black gunpowder are in the following proportion: saltpetre 75 per cent., charcoal 15 per cent., sulphur 10 per cent., and this is the proportion generally followed by English and the best foreign makers. In Norway and Sweden less sulphur is used. In France, Prussia, Russia, Spain, and America more sulphur is used.

The explosive force is generated by the saltpetre and charcoal; the sulphur raises the temperature of the freed gases, and adds to their volume by its own decomposition. The speed with which a charge of powder explodes is due to the size and density of grain. This question of grain is of the first importance to the sportsman. Annexed are representations of the more usual grains.

For a muzzle-loader, No. 3 grain has been long preferred, as the smallness of the grains permits of the powder freely entering the breech and nipple; with the breech-loader, this necessity no longer existing, the grain which gives the best results at the target is chosen.

The author, after many thousands of experiments, is of opinion that the No. 4 Alliance is unequalled for all-round shooting in guns of any gauge. For 12 and smaller bores, when strong shooting is required, the No. 4, having a larger proportion of the finer grains than is usually found in ordinary No. 4, will be found to give satisfactory results. This powder would not be obtained by mixing No. 3 and No. 4 together in *any* proportions, but would require to be specially sifted. The Basket coarse grain of Curtis and Harvey is much liked by some sportsmen, but the author has found a larger-grained powder to give better results. Even for 8 and 4 bores the author has found Pigou, Wilks, and Laurence's No. 4 Alliance more suitable than the larger-grained powders sold as best for these large shoulder guns.

To sum up the question of grain in one sentence: whilst No. 6 gives regular and even patterns, it has not the velocity of smaller-grained powders; the No. 3, 2, and 5 grained "Basket" cause the pellets to scatter more rapidly than the larger grained powders; the No. 4, therefore, is the happy medium for the ordinary purposes of the field and trap shooter.

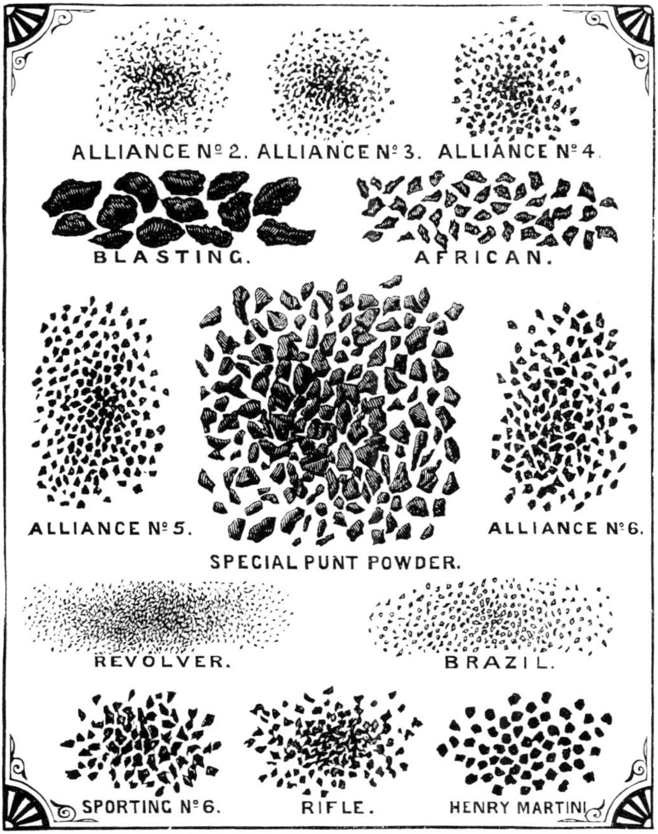

Pigou, Wilks, and Laurence's Black Gunpowders.

Messrs. Hall have introduced a mixed grain powder, which is supposed to combine quickness of ignition with quickness of combustion. The author has not found this powder to give results in any way superior to those obtained with the No. 4 Alliance.

The favourite grains in different countries vary with the locality; but the illustrations below represent fairly the grain with which a sportsman will be served—unless he specify a powder or grain—in France, Germany, or the United States.

The grains of the powder should be angular, round or cubic grains being slower to ignite, and slower in combustion. The round grain gunpowder of the German manufacturers (Cramer & Bucholz) the author has tried, and the results in all tests have shown to a disadvantage when compared with the best angular-grained gunpowders. The best quality of black gunpowder is not always the cleanest or brightest in appearance. The following ready tests may be applied when powder of a doubtful brand is about to be purchased:—

Rub a few grains in the palm of the hand, or between the finger and thumb. If it is reduced to dust with little pressure, its density is insufficient,

Foreign Gunpowders.

and the quality poor; if the colour of the dust is of inky blackness, the charcoal is of inferior quality, and the powder poor in consequence. Good gunpowder can scarcely be reduced to dust by rubbing in the hand; the dust will be of a rich dark brown, and if a little powder be ignited in a piece of paper, no residuum should be left, nor the paper burnt through.

The best gunpowder requires the very best alder-wood or dog-wood charcoal, refined sulphur, and the purest saltpetre. These three ingredients must be *thoroughly* corporated, and the powder submitted to great hydraulic pressure, before being broken up and granulated.

The charcoal of inferior quality, or badly burnt, or of the wrong wood, will spoil the quality of the powder, whatever pains be taken with the subsequent stages of manufacture.

The brown, or cocoa powders, recently introduced, owe their colour to a different treatment of the charcoal, and these powders have no special quality to recommend them for sporting purposes.

Black gunpowder will ignite when the temperature is raised to 600° Fahr.—some kinds at a lower temperature, and when exploded, will generate 6,400 atmospheres.

Nitro-compounds differ only from each other in the "base" used, and the finish of the powder. In Schultze gunpowder, pulped wood is treated with the acids, then purified, granulated, and waterproofed. In "E.C." gunpowder, cotton-waste pulped is treated with the acids, then purified, granulated, and waterproofed. In gun-cotton, cotton is the basis, and the granulating and waterproofing processes are not performed.

Schultze gunpowder is of a pale yellow tint, "E.C." is pink, or pale orange, and Dittmar and Coopal's yellowish-white.

Schultze was originally of the same appearance as the Dittmar, and the Dittmar has, the author is informed, been recently improved. Its basis is pulped wood. Coopal's is practically granulated gun-cotton. The properties of these different explosives, in comparison with black gunpowders, will be gathered from the following remarks :—

Black powder, generally speaking, has a real specific gravity of about

Schultze, Dittmar, and "E.C." Gunpowders.

1·720, whilst the Schultze powder, pressed and granulated, has a specific gravity of ·860. Therefore, a charge measuring three drams will weigh, black powder 84 grains, Schultze 42 grains. In combustion, wood powder is far more rapid than black, and therefore a greater muzzle velocity may be obtained with it than with black powder under the same conditions. It seldom gives a greater increase in velocity than five per cent., although the makers affirm that it could be made to give more if desired.

In consequence, therefore, of its rapid combustion, it is unfitted for rifles or for very large-bore guns. In small rifles of 360-bore it has been found to answer very well, but black gunpowder is in all cases more suitable where great resistance has to be overcome at the commencement.

To the great difference in the density of wood and black powder may be traced the disparity between the solid residues of the respective explosives.

Wood powders possess various advantages over black, the chief being the absence of smoke after the discharge, and the small amount of residue deposited in the barrel. This is on account of the greater percentage of available gases contained in nitro-compounds to that of gunpowder. Black gunpowders usually give about 65 per cent. solid residue and 35 per cent. available gases, which of course have to drive out of the barrel the solid residue, in addition to the charge of shot and wads in front of it ; the major portion of the solids being in a state of fine division or smoke. The best wood powder will give about 30 per cent. solid residue, 70 per cent. available gases; consequently, one-half the charge of powder by weight is equivalent in force to a full charge of black powder. This leaves, therefore, only about 15 per cent. solid residue to be expelled from the barrel, against nearly 65 parts solids from black. The solids resulting from the wood powder are expelled in a coherent form instead of smoke, thus slightly lessening the recoil.

Unconfined wood powder, in common with other nitro-compounds, may be ignited without obtaining a third of the available explosive force; to get the best results the ignition must be made by a detonator. The detonating powder contained in the ordinary sporting cap is sufficient for a sporting charge.

The "E.C." powder has many properties in common with the Schultze. Its specific gravity is about the same; the amount of solid residue left in the gun-barrel is, if anything, less; the smoke is less dense even than from Schultze, and the barrels do not heat so rapidly, and, strange to say, invariably heat from the muzzle to the breech, instead of from breech to muzzle, as is usual with black powder. Theoretically, nitro-cellulose is superior to nitro-lignum. In actual practice there is little difference.

Nitro-cellulose contains about 14 per cent. of nitroxyl (NO_2) when at its full strength; but by using weaker nitric acid in the solution, a less percentage results, and ignition by detonation will be more difficult, combustion slower, and the explosion less violent.

Black gunpowder, on an average, will fire at a temperature of 539° Fahr., whilst nitro-cellulose, or "E.C." and Schultze gunpowders, fire at 370° Fahr.

The result of heat before ignition to various explosives is attended by very different results. *Nitro-glycerine* will explode with a modicum of violence when at 60° Fahr., much more strongly at 100° and 350°, and increases in violence up to 750°, but at and beyond 750° it becomes comparatively weak, and its explosiveness is more and more feeble as the temperature is raised.

Black gunpowder is much more violent if heated to 212° before detonating, and the strength of "E.C." and Schultze powders increases in a greater ratio than black, and when heated requires less detonation.

The strength of nitro-compounds generally is better developed when the detonator is in actual contact with the explosive. The flash alone of an explosive cap would not develop nearly so much energy from the powder as would a detonator fired in the middle of the charge; but the explosion would be stronger than if the charge were fired by insertion of a heated wire, or by the application of a flame.

All nitro-compounds are more violent in their action the more tightly they are confined, and the stronger the detonation by which they are exploded.

The explosion by means of a Bickford fuse of various explosives in lead cylinders 4×8, and bores 1×4 inches, capacity of 60 cubic centimetres, resulted as follows:—1 oz. of Curtis and Harvey's No. 4 Diamond Grain increased capacity to 280 cubic centimetres; whilst ½ oz. of "E.C." increased it to 210, and a like charge of Schultze made the same increase.

The chemical action of the residue left in the barrel after firing "E.C." or Schultze is not more deleterious than that left after firing the best black gunpowders, and no more cleaning or preparation of the barrel is required with one explosive than with another.

The trials hitherto made with the new safety powder do not show that the powder equals in strength or regularity the best brands of ordinary powder, and whatever merit the powder may possess as a safety explosive it is, for sporting purposes, inferior to the explosives to which the leading sportsmen have been accustomed.

Silatoir, or "strength-producer," is the name of a Russian nitro-compound which is manufactured from a wood pulp, as is the Schultze. The pulp is boiled with nitrate of potash, and treated with sulphuric and nitric acids. This explosive has a perfect combustion, leaving absolutely no fouling in the barrel, and is quite smokeless. At the St. Petersburg gun trials of 1887 it was tried in the author's winning gun, but did not register patterns or force equal to those registered by Schultze powder in the same gun. At later trials the powder in a compressed form was tried, and penetration and patterns equal to those made with the sporting charges of Schultze powder were obtained. This powder, when granulated and perfected in manufacture, may prove a formidable rival to the favourite nitro-compounds of the day.

SMOKELESS "S.S." POWDER.

The explosives manufactured by the Smokeless Explosives Company at Ware differ from the Schultze and "E.C." in composition, for it is contended that neither nitro-cellulose nor nitro-lignum forms the base from which the powder is produced. From trials made by the author, as well as from reports he has received from sportsmen who have submitted it to practical tests, the "S.S." powder would seem to offer no particular advantage over the explosives already established. It cannot be denied that excellent results, both at the target and in the field, have been obtained with "S.S.;" nevertheless, the advantages which have been gained by its makers are rather of a commercial and technical nature than of superior qualities or powers in the explosive itself.

SECTION II.—SHOT.

LEAD SHOT is of two kinds: that which is moulded, as large buckshot, and that which is "dropped," as the ordinary small shot.

Drop shot should be made of lead without a tin alloy; it may be hardened by the patented process of the Newcastle Chilled Shot Company. Hard shot is preferable to soft for all purposes, but it is more expensive. The hardest, heaviest, and roundest shot made in England is that manufactured at Gateshead by the Newcastle Chilled Shot Company. The hardest lead shot will not injure a gun-barrel, even if the barrel be of soft Belgian metal.

The absurd notions current respecting the qualities of chilled shot scarcely demand serious consideration. Sportsmen who can treat otherwise than jokingly the assertion that shot will glance off the feathers of a bird, are not likely to dip deeply into the theories of the flight of projectiles. It is well known that several shots of a charge will sometimes take a very erratic flight, but it cannot be shown that chilled shot are more apt to do so than soft. Could these erratic pellets be examined after flight, they would be found to be of irregular shape.

The nearer to a perfect sphere each pellet of a charge is when the charge leaves the muzzle, the nearer perfect will be the flight of that charge. In passing through the barrel the shot, by pressing against one another and the barrel, become deformed, unless they are of hardened metal. Chilled shot will improve the shooting of any gun; it does not lead so much as the softer shot, and if made as it is by the Newcastle Chilled Shot Company is of the same specific gravity and free from poison as soft shot, whilst as a projectile it is superior in every way.

To the teeth either soft or chilled is equally disagreeable; it would be better for shooting could shot be made still harder than it now is.

To make shot, the lead, when molten, is poured through a sieve which has a tremulous motion conveyed to it by a geared machine; this motion causes the cooling lead to form itself into small globules, and these globules retain their shape as they fall down the shaft into the water placed for them. A large percentage of these globules are not perfectly spherical. From the best quality of shot *all* ill-shapen pellets are removed, and care is taken that the shots of one nominal size are all as nearly as possible the same size. Unless this were so, the shooting would be inferior; no good shooting can be made by mixing shots of two different sizes in the one charge.

There is in commoner shot a great diversity in the shape and size of the individual pellets, and the average size will not be found to correspond with the published size.

American and Continental shot is not so regular, either in size or shape, as the English; and there are frequent discrepancies between the printed list of sizes and the actual dimensions of the shot. In America, as in England, there seems to be no association of manufacturers for the purpose of producing shot of standard gauges; and as it is impossible to determine the merit of a gun's performance at the target without knowing the number of pellets contained in the charge, the following tables of sizes of the leading manufacturers may be found useful:—

STANDARD SIZES OF ENGLISH SHOT.

LANE AND NESHAM, LONDON.		NEWCASTLE CHILLED SHOT CO., GATESHEAD-ON-TYNE.	
Size.	No. of Pellets to the Ounce.	Size.	No. of Pellets to the Ounce.
A A A A	30	A A A	40
A A A	35 to 40	A A	48
A A	40	A	56
A	45	B B B B	56
B B B	50	B B B	64
B B	58	B B	76
B	75	B	88
1	80	1	104
2	112 to 120	2	122
3	135	3	140
4	175 to 180	4	172
5	218 to 225	5	218
6	278 to 290	6	270
7	340	6*	300
8	462	7	340
9	568	8	450
10	985	9	580
Dust	1672	10	850
*S G	11	11	1040
*S S G	15	12	1250
*S S S G	17	Large Dust	1700
*L G	5½	Small Dust	2800
M G	9	S G	8
—	—	S S G	11
—	—	S S S G	14

* Walker, Parker, & Co. London sizes.

SIZES OF AMERICAN DROP SHOT.

T. N. Sparks, Philadelphia.		Tatham, Bros. New York.		N.Y. Lead Co., New York.		Le Roy Co., New York.		St. L. Shot Co. St. Louis, Mo.		Selby Co., San Francisco.		Chicago Co., Chicago, Ill.		Merchant's Co., Baltimore.		Dubuque Co., Dubuque, Iowa.	
Size.	Pellets	Size.	Pellets	Size.	Pellets	Size.	Pellets	Size.	Pellets	Size.	Pellets	Size.	Pellets	Size.	Pellets.	Size.	Pellets
F	22	F F	24	—	—	T T T T	24	000	29	—	—	—	—	T T T T	22	—	—
T T	36	F	27	—	—	T T T	27	00	32	—	—	000	27	T T T	26	000	27
T	41	Y Y	31	T T	32	T T	32	0	40	—	—	00	33	T T	30	00	33
B B B	48	T	36	T	38	T	38	B B B	46	—	—	0	38	T	34	0	38
B B	55	B B B	42	B B B	44	B B B	44	B B	53	B B B	46	B B B	46	B B B	39	B B B	46
B	63	B B	50	B B	49	B B	49	B	63	B B	55	B B	53	B B	45	B B	53
1	80	B	59	B	58	B	58	—	—	B	68	B	62	B	52	B	62
2	90	1	71	1	69	1	69	1	79	1	79	1	75	1	60	1	75
3	118	2	86	2	82	2	82	2	98	2	89	2	92	2	77	2	92
4	130	3	106	3	98	3	98	3	116	3	118	3	118	3	94	3	118
5	182	4	132	4	121	4	121	4	163	4	134	4	146	4	115	4	146
6	245	5	168	5	149	5	166	5	181	5	170	5	172	5	140	5	172
7	305	6	218	6	209	6	209	6	252	6	215	6	246	6	180	6	246
8	426	7	291	7	278	7	278	7	306	7	303	7	323	7	225	7	323
9	615	8	399	8	375	8	375	8	426	8	420	8	434	8	365	8	434
10	950	9	568	9	560	9	560	9	584	9	592	9	596	9	610	9	596
11	1660	10	848	10	822	10	822	10	981	10	874	10	854	10	1130	10	854
12	3316	11	1346	11	982	11	982	11	1603	11	1404	11	1414	11	2200	11	1414
Dust	5910	12	2326	12	1778	12	1778	12	2305	12	2030	12	2400	12	3200	12	2400
—	—	—	—	—	—	—	—	—	—	—	—	—	—	13	12200	—	—

CONTINENTAL SIZES.

French.						Italian.		Belgian		Austrian.		Prussian.	
Paris.		Lyons.		Marseilles.		Genoa.		Brussels.		Ramer & Co.		Cologne.	
Size	Pellets.	Size.	Pellets	Size.	Pellets	Size.	Pellets	Size.	Pellets	Size.	Pellets.	Size.	Pellets
000	40	0000	28	0000000	28	00000	49	0000	23	0000	—	0	2 to 3
00	50	000	30	000000	33	0000	53	000	33	000	—	00	3 to 4
0	60	00	33	00000	35	000	54	00	39	00	21	000	4 to 5
1	74	0	34	0000	39	00	64	0	45	0	23	0000	5 to 6
2	81	1	37	000	49	0	87	1	56	1	26	A A A	19
3	90	2	43	00	54	1	102	2	69	2	—	A A	26
4	108	3	53	0	62	2	141	3	79	3	35	A	29
5	189	4	81	1	71	3	146	4	128	4	41	B B	38
6	214	5	86	2	75	4	156	5	169	5	—	B	43
7	262	6	178	3	97	5	162	6	216	6	59	1	50
8	337	7	184	4	114	6	193	7	268	7	69	2	64
9	686	8	261	5	144	7	268	8	—	8	82	3	82
10	988	9	443	6	150	8	308	9	—	9	92	4	97
11	—	10	721	7	190	9	395	10	1052	10	124	5	123
12	—	11	1254	8	229	10	527	11	—	11	155	6	167
—	—	12	2218	9	312	11	714	12	—	12	203	7	230
—	—	—	—	10	485	12	1114	—	—	13	283	8	336
—	—	—	—	11	920	13	2278	—	—	14	303	9	442
—	—	—	—	12	1177	—	—	—	—	15	652	10	696
—	—	—	—	—	—	—	—	—	—	16	809	11	1223

The pellets are counted to the English ounce, which equals 28 grammes.

SECTION III.—CARTRIDGE CASES AND WADDING.

CARTRIDGE CASES.

THE chief requisites in cartridge cases for shot guns are: perfect ignition, uniformity of size, capability of resisting atmospheric changes and moderate pressure.

Nothing is more tantalising to the sportsman than miss-fires; hang-fires, too, are a great nuisance. To enjoy shooting, there must be perfect freedom from tight-fitting cartridge cases, and the cases must be well made to stand heavy charges and extract freely. Not unfre-

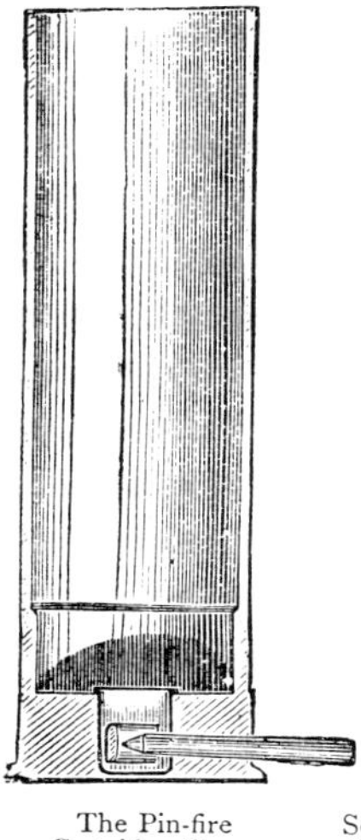

The Pin-fire Cartridge Case.

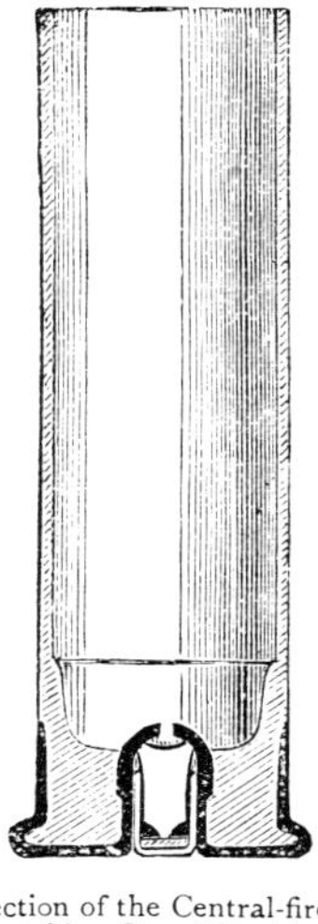

Section of the Central-fire Cartridge Case (Eley's).

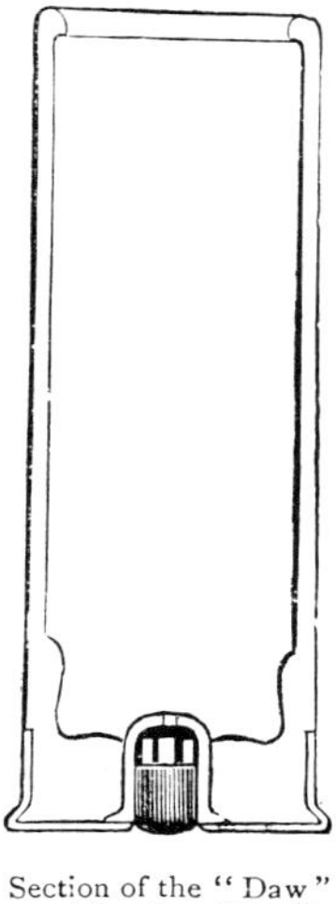

Section of the "Daw" Central-fire Cartridge Case.

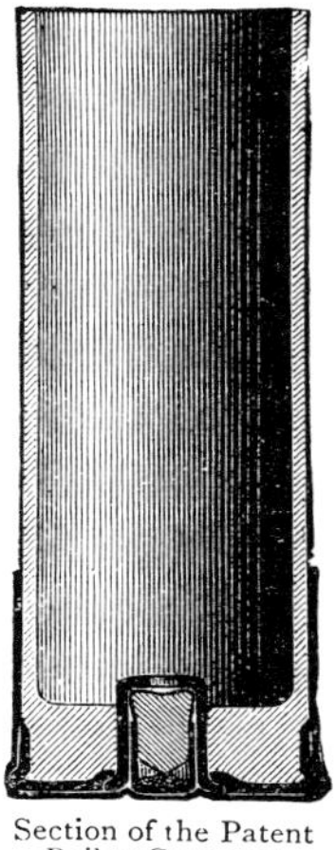

Section of the Patent Bailey Case, and View of Gas-tight Base.

quently with cheap cases the heads are pulled off by the extractor, the cylinder being left in the chamber to be removed by the *grip extractor*, sold by gunmakers for this purpose, or cut out with the shooter's knife.

The cases must be of exact size also in the rim; many have small rims, so that in loading they slip past the extractor, and prevent the gun being closed. It is also of importance that the case be substantially made, capable of resisting such pressure as careless packing in the magazine, or the usual accidents of travel may occasion. A stout, well-made case will keep the powder in better condition than will a thin, unglazed, imperfectly finished case.

There are several good makers of paper cases in the market. Eley's green, as made by this celebrated firm for the last twenty-five years, give general satisfaction. The blue case may be used for a different charge as a distinctive colour; in quality it is but little inferior to the green. The buff and red cases sold for Schultze and "E.C." powder serve this same purpose. The cap in red cases for "E.C." powder is somewhat stronger than that generally used by Messrs. Eley, and ensures the quick ignition of this powder.

Stronger cases than any made by Eley are the "High Life" and

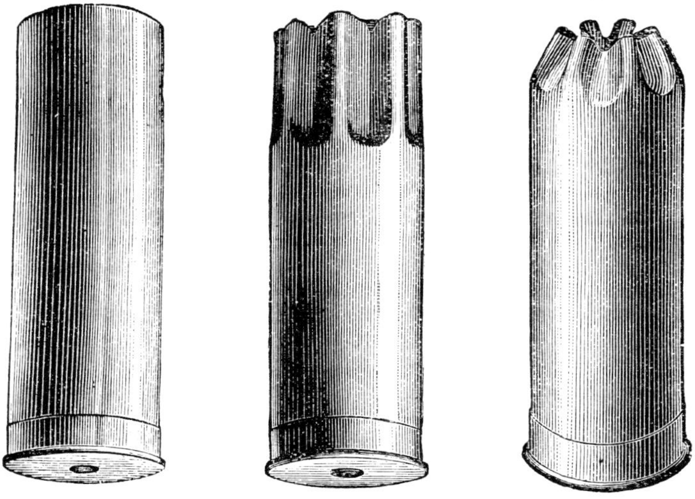

Kynoch's "Perfect" Case. | Method of Crimping the Kynoch "Perfect" Case with Greener's Corrugator. | The "Perfect" Case Crimped and Compressed.

"Demi-High Life" cases, much used for pigeon shooting. They have a coil of thin brass foil outside the paper, strengthening the case near the base, and making the extraction more easy. They may be used advantageously with heavy charges, or in guns with weak or much-worn breech-actions.

Messrs. Kynoch, Limited, have a "perfectly gas-tight" cartridge case, in which a solid drawn cap extends from the base to the middle of the case outside the paper, which is superior to a metal lining, and renders the case as stable as the "Demi-High Life."

Messrs. Joyce have a paper case with a metal capsule covering the cap, to prevent the escape of any gas into the breech mechanism.

The thin brass perfect case of Messrs. Kynoch has greater capacity than the paper cases of the same nominal gauge. These cases are the same price as the best paper cases, and as they do not swell, and keep the powder in good condition, they are much favoured by some sportsmen; others do not like the "feel" of the sharp edges of the metallic cases, and continue to use paper. The latest pattern of perfect case has the base filled with paper pulp, which strengthens the rim, renders the base solid, and lessens the liability to miss-fire. These cases do not give way in the least; and if the primers are kept of uniform good quality sportsmen may congratulate themselves upon having a truly "perfect" case. In America paper cases, or "shells," are dearer than in England, and not equal in quality to those made here. The American metallic cases are good, but very dear. On the Continent a very cheap paper case—colour greyish-brown—is sold, which is of such poor quality that all sportsmen would do well to avoid using them.

WADDING.

The wadding used in the shot gun is of three varieties: 1st, the simple cardboard wad; 2nd, a felt wad; 3rd, a hard felt paper-faced wad, known as the "pink edge" or the field wad. A cardboard or waterproof (pitch paper) wad must be placed over the powder; this must be followed by a lubricating felt wad, usually 3-8ths of an inch in thickness. The top wad over the shot must be of the thinnest cardboard. Pink-edged, pink-faced, "field," and thick cardboard wads, cloth wads, and black wads are used for special purposes, as specified in the section on cartridge loading.

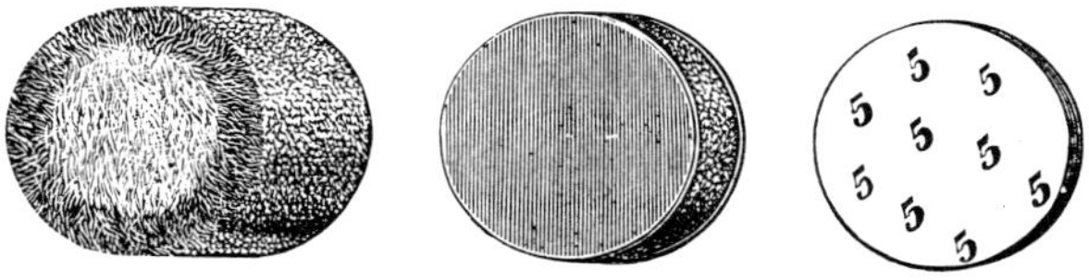

Shot Wads.

Some special wads have been introduced for the purpose of increasing the range of the gun; one which is much used on the Continent consists of a conical cup wad, like the improved Lancaster Concentrator reversed, and its use in cylinder guns is advantageous; in choke-bores it should not be used.

Another wad is so constructed that immediately it leaves the muzzle of the gun it falls to dust, and this wad, if fixed over the shot, slightly increases the pattern at short ranges; if over the powder without other wadding, it reduces both pattern and penetration. One firm of English gunmakers use an oily preparation of soft wood sawdust for wadding.

In the United States a metal top wad is sometimes used to *fix* the charge in paper or brass cases ; it is known as the " Star " wad. The author has no experience of its advantages.

The foregoing illustrations of wads generally used in England will enable the tyro to recognise them, especially if the following particulars are remembered.

The best felt wads are elastic, of a light pink colour, deeper at the greased edge. Cheaper qualities are of a deeper tint, and the plainest are brown in colour and hard and close in texture. The " Field " wad is black (pitched paper) on one face, pink paper on the other ; the edge is greased. Pink-edged wads are greased at the edge, and have paper faces of a light pink tint ; they should be of elastic felt.

LOADED CARTRIDGES.

A few hints and theories concerning the choice and loading of cartridges may be welcome.

The " Life " cases made in Belgium have large caps, and give perfect ignition to any explosive used for sporting purposes. The fulminate is of a different kind to that found in cases of English make. The bases are stronger and more solid, and they rarely, if ever, burst at the rim. These cases have given great satisfaction, and as they may be more readily de-capped than those made on other systems, they enjoy an enviable reputation amongst sportsmen who reload their cartridge cases. The large cap prevents the escape of gas into the striker hole, and this is but one of their several other advantages.

Felt wadding *must* be used between the powder and shot, a wad $\frac{3}{16}$ inch thick is enough for a 28-bore, and $\frac{3}{4}$ inch would not be too thick for an 8-bore ; as there is not a wad so thick, a pink-edged wad must be used as well ; $\frac{3}{8}$ or $\frac{7}{16}$ is the correct thickness for a 12-bore wad.

Wads need not fit the case tightly. There is little doubt but the wads are expanded in the cartridge case before the shot is started. The felt wad serves to clean the fouling in the barrel left from the firing of the previous charge. The blackened edges of the felt wad are not caused by the gases of the ignited explosive singeing the wad, but the fouling in the barrel, which fouling also hardens the wad edge very materially. It is quite possible that this fouling is in some way beneficial to the gun barrel, as it acts as a lubricant for the shot, and it is well known that a shot barrel with a certain amount of fouling will throw the shot closer and harder than a dry, bright barrel.

Large charges of black powder, and all charges of Schultze or " E.C.," should be wadded with a pink-edged as well as the felt wad, as the shooting is thereby improved. In the same manner, if a lighter charge of shot is used, extra wadding must be employed. If the thick felt wad is soft and greasy, it may sometimes be picked up after firing with several pellets of the shot charge buried in or adhering to it ; with four wads it is therefore necessary to put a card wad over the felt.

TECHNICALITIES.

The wads when fired fall a little to one or the other side of the line of fire, and at a short distance from the muzzle of the gun.

If the time occupied from the instant of pulling the trigger to the pellets of the charge striking the object at 35 yards' distance be ·1237 sec., which is about the average when using the standard charge in a 12-bore gun, the periods may be divided as follows:—Time occupied from pulling of trigger to striker touching the cap, ·0022 sec.; cap to muzzle of gun, ·0055 sec.; and muzzle to object, ·1160 sec., which is a mean velocity for

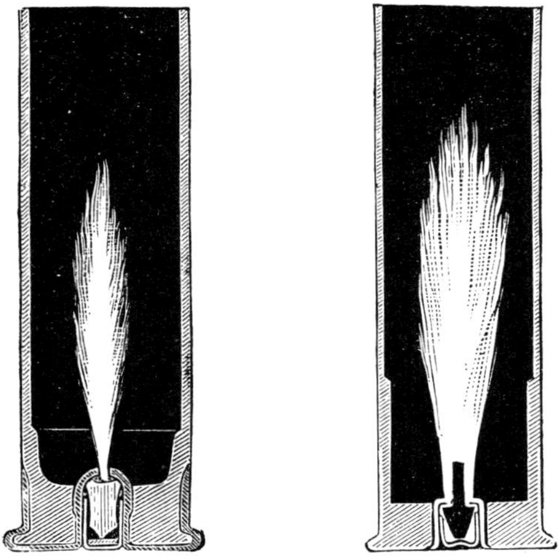

"Eley" Case. "Life" Case.
Ignition of Central-fire Cases.

View of the Cap-chambers and Section of Caps and Anvils of the "Life" and "Eley" Cases.

the 35 yards of 905 feet. With the larger cap and Schultze powder the mean velocity will be 910 feet per second, but the total time will be ·1235 sec., as, although the time occupied by the ignition and combustion of the Schultze is ·0004 more than that required by No. 4 black powder, the force generated throws the pellets the 35 yards in ·0006 less time than the black. In Schultze gunpowder the slowness of ignition is compensated or by the greater amount of gas liberated by the combustion; and if the ignition of the powder be quickened by the use of a large cap and more fulminate the pressure exerted by this larger body of gas is such as to cause the pellets it projects to travel at a greater speed than it is possible to project them with any other explosive used in shot guns.

A mixture of various-sized angular grains of black gunpowder of the best quality gives the nearest approach to the results obtained from Schultze gunpowder.

It is generally admitted that the stronger the ignition of the explosive the better the shooting. Messrs. Eley have, in their ordinary central-fire cap, an article which may be absolutely relied upon for freedom from miss- and hang-fires. That they were loth to depart from their standard pattern is not surprising, but certain of the nitro-compounds now in use requiring such, they produced a cap charged more heavily with fulminate, but otherwise of the same shape and style as their usual primer.

The cut shows the exact size of the orifice through which the flash must pass, both in the Eley case 1, and the "Life" case 2; it will be seen that there is a wide difference.

The cap of the "Life" case 2 is larger, and the anvil of a different shape.

The flash with the "Life" cases is very full and fills the case; from the smaller cap, with the cap chamber pierced with a small round hole only, the flash is much smaller in volume, as well as being of inferior force.

CARTRIDGES.

The large cap of the *Sporting Life* cartridge case ensures the perfect ignition of the explosive, producing an increased velocity, greater penetration, and larger range. The size and form of the flash-hole, the shape and arrangement of the heavily charged cap, and the broad anvil render miss or hang-fires impossible, and produce perfect combustion of large or small charges of all explosives; the increased diameter of the cap prevents fouling through the striker hole or erosion around it.

Another case of Kynoch's make is similar to the perfectly gastight, but the brass is extended to within one quarter inch of the mouth of the case. It is called the "grouse" case, and the advantage claimed for it over other paper cases include non-liability to change when exposed in damp places or used in wet weather.

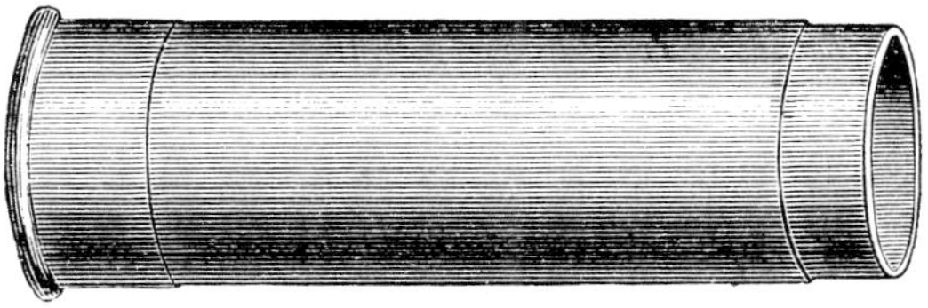

Messrs. Kynoch's "Grouse" Cartridge Case.

SECTION IV.—GUN CASES AND IMPLEMENTS.

THE gun that is worth owning is worth preserving. If kept in a gun rack, or better still, a dust-proof gun cupboard, it will last longer, and if put away clean will always be ready for use. To take guns from place to place, a case is necessary; if they are to be sent, a substantial oak case, leather-covered, is the best—such a case well made is worth about £5. It affords complete protection to the gun, and will itself withstand the roughest usage.

Sole leather cases—that is to say, cases in which best leather is *sewn* to pine frames—are light and handy, and do well to carry guns in, but they must not be used as packing-cases; and although they will upon occasion stand several trips to the Rocky Mountains, they are not adapted to the rough usage they receive in the goods van, and do not protect the gun as will the oak-covered case. A best sole leather case is worth £4. The leg-of-mutton case affected by trap-shooters does not give much protection to the gun, but it is very light, and serves well to carry the gun; the cost is £3. A case to carry the gun at full length has been recommended, but its bulk makes it inconvenient in many situations where the shorter case is no encumbrance.

It is preferable to carry cartridges in a separate magazine than in a tray in the gun cases. These magazines are made to carry 50—the neat little

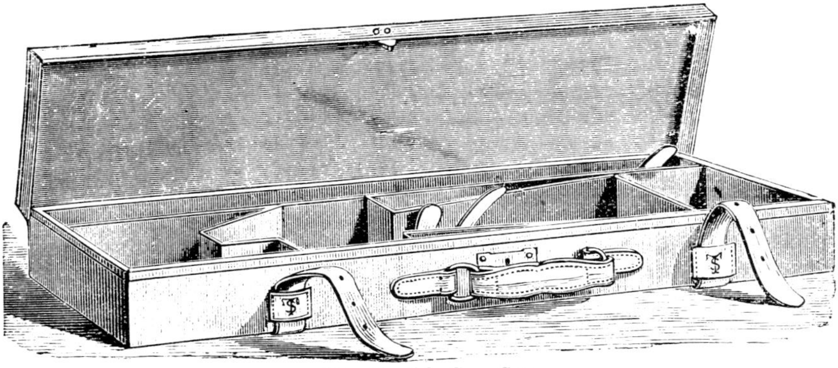

The English Gun Case.

case carried by the trap-shooter—and 100, 200, 300, 400, or 500, the last a substantial trunk, heavily made, and able to withstand luggage porters' care-less handling.

The divisions are preferable as in the English magazine, for the cases

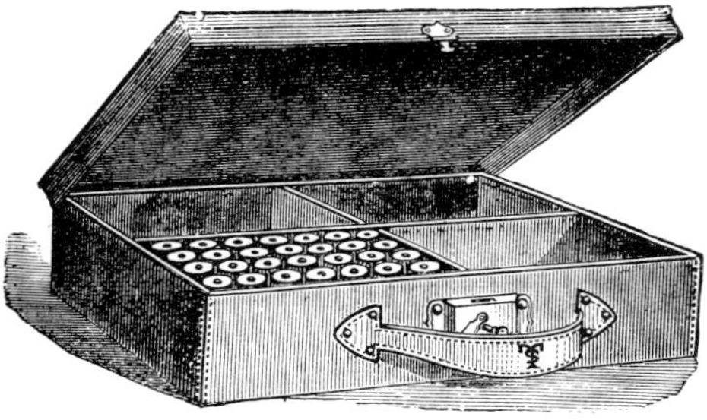

American Cartridge Magazine.

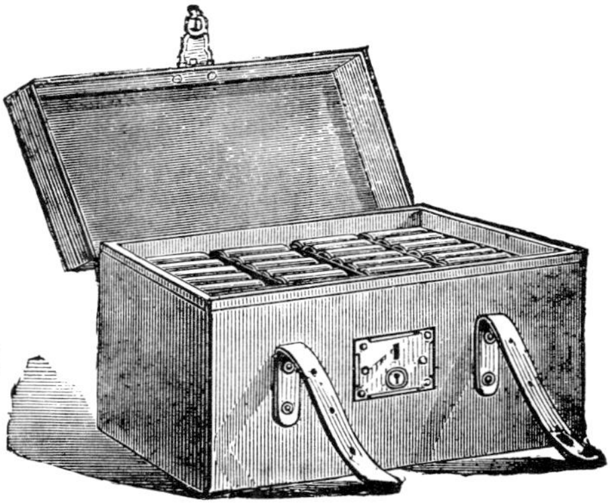

English Cartridge Magazine.

then travel better, and are more easily packed into and removed from the magazine.

The impedimenta in England may be restricted to a full set of cleaning implements, including pocket cleaner and chamber brushes, action brush, tow, rag, flannel, and oil. A pocket extractor is useful, and a pair of turn-screws may be kept in the cases.

Cartridge loading apparatus will be found useless in England, where ammunition is cheap, and it is not the rule to reload cases.

When travelling abroad, powder and shot measures should be taken, also a rammer and turnover de- and re-capper and supply of caps, gun-powder, wadding, and a little shot.

Cartridges are best carried in a magazine of solid leather, or wood canvas-covered. In the field cartridges may be carried in the pocket. Cartridge belts cannot be recommended for ordinary shooting, although there are times when they are very useful, if not indispensable. Cartridge bags to hold 75 or 100 cartridges are large enough for all purposes, and will be found to hang very heavily if much walking is done. Two small bags feel much lighter than one large one. The Quellan patent cartridge carrier is very handy, but is open to the same objections as the bag, viz., the weight always bearing upon the one shoulder tires more quickly than if the load be distributed in the pockets of a shooting jacket.

PART III.—TRAP SHOOTING

SECTION I.—LIVE BIRD SHOOTING.

THE shooting of live birds is supposed to be a cruel sport, whereas, as carried on at the best English and foreign clubs, it is a perfectly legitimate pastime, and as free from cruelty as any shooting at birds or beasts can be.

Notwithstanding that the opponents of pigeon shooting allege, with some truth, that pigeon shooting, unlike British field sports, is of lowly origin, it may be said that for several generations it has been a favourite pastime with many of the best known sportsmen of this country, as well as patronised of late years in all parts of the world by thorough sportsmen.

As a means of ascertaining the relative skill of "wing shots," or as a method of determining individual prowess with the shot gun, the shooting at pigeons freed from a trap offers advantages over any other method, and this fact alone would make pigeon shooting a favourite pastime with those who consider themselves good game shots.

On the other hand, it must be allowed that pigeon shooting *may* have originated with gamekeepers and poachers, who found therein a legal method of determining who was the best shot. Pigeon shooting has long been a holiday pastime with the frequenters of low public-houses, and has been and still is used as a means by sharpers to fleece the unwary young sportsman. It is wise to shoot pigeons at recognised clubs only, whether in England or elsewhere; and before accepting an invitation to shoot a friendly match, to make sure of the character of the person who invites the contest, or experience at the trap may be very dearly bought.

Pigeon shooting as a sport may be said to date from about the middle of this century, although there were occasional matches and contests earlier. The first handicap is said to have been shot upon Mr. Purdey's grounds at Willesden in 1856, but previous to this there had been fashionable contests at the "Old Hats" public-house, on the Uxbridge Road at Ealing, near London. The "Old Hats" obtained its name from the fact that the pigeons used for the matches were placed in holes in the ground, and were covered with old hats. The "Red House" at Battersea was afterwards the favourite metropolitan resort for wager shooting. The first *boná-fide* Pigeon Club was formed at Hornsey Wood House. Traps were

Pigeon Shooting at Hornsey Wood.

used here, and the "small cannon" which were formerly in vogue as pigeon guns were discountenanced, and the ordinary double-barrelled fowling-piece substituted. The late J. H. Walsh (Stonehenge) gave a sketch of this fashionable ground in his book on the gun, which illustration is, by the kind permission of his publishers, reproduced here. Since the founding of the Hornsey Wood Club numerous clubs have come into existence, and amongst the most celebrated in England are :—

THE GUN CLUB, Notting Hill, London. Secretary, Mr. G. A. Battock, 4, Carlton Street, S.W.

Winter shooting commences the first week in November, and is continued every Saturday until March.

The summer shooting commences in April, and is continued twice or thrice weekly until July.

THE HURLINGHAM GUN AND POLO CLUB, Fulham, London. Secretary, the Hon. D. Monson, Hurlingham.

Summer shooting commences in April or first week in May, and is continued twice or thrice weekly through the summer.

The International week of the Hurlingham and the Gun Clubs is generally fixed for the 15th of July, and the combined sweepstakes money and other prizes offered by the two clubs to be shot for in these contests are worth about £3,700.

THE INTERNATIONAL GUN AND POLO CLUB, Brighton. Secretary, Mr. Holt, 173, Piccadilly, London, W.

Frequent meetings during the summer, and occasional shooting in the winter, with additional fixtures for the week following the International week at London.

THE UNION GUN CLUB, "The Welsh Harp," Hendon. Secretary, Mr. Maas, 1, Dover Street, Piccadilly.

Frequent matches and handicaps on Saturdays.

THE MANCHESTER GUN CLUB. Secretary, Mr. Ainsworth, Barton Arcade, Manchester.

THE WOLVERHAMPTON GUN CLUB. Secretary, Mr. Sweetam, Star Hotel, Wolverhampton.

THE OLDHAM GENTLEMEN'S GUN CLUB. Secretary, Mr. Hitchen, White Lion Hotel, Oldham.

THE WITTON GUN CLUB. Secretary, Mr. Twiss, Witton Road, Witton, Birmingham.

There are also in or near London *The Commercial Gun Club*, *The Licensed Victuallers' Gun Club*, *The Wimbledon Gun Club*, and *The Mid-Kent Gun Club*. There is also pigeon shooting at the *Ranelagh* and at the *Orleans Clubs* by arrangement.

In the provinces there are good clubs at Cheltenham, Maidenhead, Redcar, Barrow, Taunton, Exeter, Chester, Newton-le-Willows, and Stockton-on-Tees.

The London Gun Club.

Abroad.

FRANCE.—*Cercle des Patineurs*, Bois de Boulogne, Paris. Secretary, M. Payme

MONTE CARLO—principality of Monaco. Secretary, M. Blondin; address: Monte Carlo. The season commences about the 10th of December, and there is shooting every Tuesday, Thursday, and Saturday until the middle of January, when the International Contests commence. At the International Contests there are, in addition to the sweepstakes, valuable pieces of plate—in the "*Prix d'ouverture*" the plate is usually worth about £80; in the "*Grand Prix du Casino*," £800, with about £280 as share of the entrance-money; the second taking £160, and one-fourth of the entrance-money, the third £80 and one-fourth, and the fourth £40 and 15% of the entrance-money. The *Prix de Monte Carlo*—a prize of plate and £120 joined to a sweepstakes. *Prix de Consolation*—a piece of plate and £40 joined to a share of the entrance-money. Every third year there is also the great contest for the "*Championnat Universel*," which, in addition to the great prestige as a pigeon shot the winner has, is worth about £400 in money.

The International Contests are followed by meetings twice weekly until April.

The Monaco Meetings are attended by the best shots of all nations, and may be participated in by members of the chief Pigeon Clubs in Europe.

BOULOGNE-SUR-MER.—Secretary, M. Blondin, Rue de Thain 6, Paris. The International Meetings held here during the last week in July, and the first in August, are attended by first-class shots of all nationalities. The prizes offered are worth £400.

There are also good meetings at Abbeville, Amiens, Bordeaux (*Saumur*), Cambrai, Havre, Lille, Lyons, Macon, Nantes, Rheims, Rouen, Tours, and Valenciennes, during the summer months. During the winter there are meetings at Arcachon, Cannes, Marseilles, Pau, and several other winter resorts.

At Paris, in addition to the *Cercle des Patineurs*, there are the *Tir d'Acheres* and the *Tir du Parc* (Saint-Ouen), where pigeon shooting is practised.

BELGIUM.—BRUSSELS. The *Tir aux Pigeons*, Bois de la Cambre, is the chief club. The Secretary is M. Brancquaert, Boitsfort, par Bruxelles, and he also acts for the Societies at *Spa*, and at *Ostend*, where there are important meetings. At *Antwerp* there is also a good club, the Secretary is M. Marchant, of List, Shootzen, Anvers.

AUSTRIA.—At Vienna, The Pigeon Shooting Club, of which Mr. Georg Guttera, of the Jockey Club, Vienna, is the Secretary. Also a large and influential club at *Trieste*.

GERMANY.—The *Union Club* of Berlin have a pigeon ground near their racecourse at Hoppegarten, and there is pigeon shooting at Ems, Baden-Baden, and Wiesbaden.

In SPAIN there are several clubs, some of which hold important International Meetings. The chief clubs are at Madrid—the late King of Spain was a member of this club and a regular attendant—at Seville, at Cordova, Cadiz, Granada, and Valencia. In PORTUGAL there are clubs at Lisbon and Oporto.

In ITALY, clubs at Rome, Milan, Florence, Genoa, Venice (a good summer meeting), Bologna, Padua, and Turin. In SICILY at Palermo, Messina, and Catania.

In RUSSIA there are two clubs at St. Petersburg, and one at Moscow. The Moscow club is a new one, but is fitted with all latest appliances, and the shooting is exceptionally good.

There are also clubs at *Pesth*, and at *Constantinople*, and at many other towns still more rarely visited by English pigeon shooters. At all of these clubs an Englishman who is a member of either of the London clubs, would be permitted to participate in the shooting. There is no pigeon shooting at live birds in Norway or Sweden, but there are clay pigeon clubs at *Christiania, Gothenburg*, and *Stockholm.*

In AMERICA gun clubs are numerous. There are more than fifty live bird gun clubs in the United States. Englishmen should procure the "Club Directory" (the Clucas Publishing Co., New York), where the leading gun, rifle, shooting, and hunting clubs will be found detailed, with the addresses of their secretaries.

There is a good club at *Buenos Ayres*, and several at *Monte Video.*

In AUSTRALASIA respectable clubs will be found in every town.

RULES OF LIVE BIRD SHOOTING.

In England, as well as in America, the English Colonies, and in most clubs founded upon the same basis as our English clubs, the Hurlingham Rules are accepted wtthout important modifications. At the *Gun Club*, however, the gun must be half a pound less than the Hurlingham limit, which is 8 lbs.

THE HURLINGHAM CLUB RULES.

1. The referee's decision shall be final.
2. The gun must not be held to the shoulder until the shooter has called "Pull." The gun must be clear below the armpit, otherwise the referee shall declare "No bird."
3. A miss-fire is no shot under any circumstances.
4. If the shooter's gun miss fire with the first barrel, and he use the second and miss, the bird is to be scored lost.
5. If the miss-fire occurs with the second barrel, the shooter having failed to kill with his first, he may claim another bird; but he must fire off the first barrel with a cap on, and a full charge of powder, before firing the second.

6. The shooter's feet shall be behind the shooting mark until after his gun is discharged. If, in the opinion of the referee, the shooter is balked by any antagonist or looker-on, or by the trapper, whether by accident or otherwise, he may be allowed another bird.
7. The shooter, when he is at his mark ready to shoot, shall give the caution "Are you ready?" to the puller, and then call "Pull." Should the trap be pulled without the word being given, the shooter may take the bird or not; but if he fires, the bird must be deemed to be taken.
8. If, on the trap being pulled, the bird does not rise, it is at the option of the shooter to take it or not; if not, he must declare it by saying, "No bird;" but should he fire after declaring, it is not to be scored for or against him.
9. Each bird must be recovered within the boundary, if required by any party interested, or it must be scored lost.
10. If a bird that has been shot at perches or settles on the top of the fence, or on any part of the buildings higher than the fence, it is to be scored a lost bird.
11. If a bird once out of the ground should return and fall dead within the boundary, it must be scored a lost bird.
12. If the shooter advances to the mark and orders the trap to be pulled, and does not shoot at the bird, or his gun is not properly loaded, or does not go off, owing to his own negligence, that bird is to be scored lost.
13. A bird shot on the ground with the first barrel is "No bird," but it may be shot on the ground with the second barrel, if it has been fired at with the first barrel while on the wing; but if the shooter misses with the first and discharges his second barrel, it is to be accounted a lost bird, in case of not falling within bounds.
14. All birds must be gathered by the dog or trapper, and no member shall have the right to gather his own bird, or to touch it with his hand or gun.
15. In Single Shooting, if more than one bird is liberated, the shooter may call "No bird," and claim another shot; but if he shoots, he must abide by the consequences.
16. The shooter must not leave the shooting mark under any pretence to follow up any bird that will not rise, nor may he return to his mark after he has once quitted it to fire his second barrel.
17. Any shooter found to have in his gun more shot than is allowed, is to be at once disqualified. Any loader supplying in sweepstakes or matches cartridges loaded in excess of the authorised charge, will be dismissed from the Club grounds.
18. None but members can shoot except on the occasion of private matches.

19. No wire cartridges or concentrators allowed, or other substance to be mixed with the shot.
20. In all handicaps, sweepstakes, or matches, the standard bore of the gun is No. 12. Members shooting with less to go in at the rate of half-a-yard for every bore less than 12 down to 16-bore. Eleven-bore guns to stand back half-a-yard from the handicap distance, and no guns over 11-bore allowed.
21. The winner of a sweepstakes of the value of ten sovereigns, including his own stake, goes back two yards; under that sum, one yard, provided there be over five shooters. Members saving or dividing in an advertised event will be handicapped accordingly.
22. Should any member kill a bird at a distance nearer than that at which he is handicapped, it shall be scored no bird, but should he miss, a lost bird.
23. $1\frac{1}{4}$ oz. of shot and 4 drs. of black powder, or its equivalent in any other description of gunpowder, is the maximum charge. Size of shot restricted to Nos. 5, 6, 7, and 8.
24. All muzzle-loaders shall be loaded with shot from the club bowls.
25. If any bird escapes through any opening in the paling, it shall be a "No bird."

RULES FOR DOUBLE RISES.

1. In Double Shooting, when more than two traps are pulled, the shooter may call "No birds," and claim two more; but if he shoots, he must abide by the consequences.
2. If, on the traps being pulled, the birds do not rise, it is at the option of the shooter to take them or not. If not, he must declare by saying "No birds."
3. If, on the traps being pulled, one bird does not rise, he cannot demand another double rise; but he must wait and take the bird when it flies.
4. A bird shot on the ground, if the other bird is missed, is a lost bird; but if the other bird is killed, the shooter may demand another two birds.
5. If the shooter's gun misses fire with the first barrel, he may demand another two birds; but if he fires his second barrel, he must abide by the consequences. If the miss-fire occurs with the second barrel, the shooter having killed with the first, he may demand another bird, but may only use one barrel; if he missed with the first barrel, Rule 5 in Single Shooting will apply.

On the Continent, especially in Southern Europe, the Rules of the

Cercle des Patineurs are adhered to. They also rule at Monaco, so they are appended in full :—

PIGEON SHOOTING RULES.

1. A pigeon shooting club is established at the Cercle des Patineurs.
2. Members of the club only are admitted, but nevertheless for persons other than members, on the request and under the express responsibility of a member, an entrance-ticket to the grounds may be obtained, available for one day. This ticket, which costs 10 francs, is not transferable, and ought to be signed by the member who requested it and a member of the committee. Members of the Hurlingham and London Gun Clubs are considered honorary members of the Cercle des Patineurs, unless their visit to France should exceed two months, in which case they ought to be formally presented.
3. The committee have the right to reserve certain days for public shooting, and to issue special tickets upon those occasions.
4. The gun-maker to the club will place at the service of members of the club guns, cartridges, and accessories. Every liberty is given to others to bring the guns and cartridges of the shooters.
5. 10-bore is the largest bore allowed, 10 grammes, 20 of powder, and 36 grammes of shot is the heaviest load permitted. No. 5 is the largest shot that may be used.

[English measure, 4 drachms powder and $1\frac{1}{4}$ of shot.]

6. Wire cartridges and cartridges specially made to increase the range are, on grounds of safety, strictly forbidden ; nevertheless, concentrators may be used.
7. The members of the committee choose one or more referees, whose decisions in all cases shall be final. They also act as handicappers. They may appoint one or several persons to form a handicap who are strangers to the committee, and if they deem it advisable they can examine the cartridges and loads, and enforce the rules.
8. The boundary within which the pigeon must fall to be scored good is the half of a circle, having a radius of 80 mètres (87 yards). In the centre is the pavilion. The distance from the centre trap to the boundary is 50 mètres (54 yards, 2 feet).
9. The traps are five in number, and five mètres (5 yards, 1 foot, 5 inches) from each other.
10. The shooter must place himself at the exact distance which has been assigned to him, and his feet must not be in advance of the line drawn to mark the distances. The gun is not to be shouldered, but the stock must be below the shoulder of the shooter. If these regulations are not obeyed, the referee may declare the shot "No bird" or "Lost" at discretion.

11. If the shooter is baffled by a competitor or spectator, or if he is discomposed through any accident whatever, he may claim another bird.
12. The shooters must succeed each other at the mark without interruption, save in the case of accident, when time will be allowed at the discretion of the referee.
13. The shooter, when at the mark and ready to shoot, should cry "Pull." Should the trap open before he says the word, it is at his option to take the bird or not; but if he shoots, the shot will be scored.
14. If the trap is sprung and the pigeon does not rise, it is at the option of the shooter to accept or refuse it.
15. If the pigeon is killed before rising it is "*No bird.*" If it is missed whilst on the ground with the first barrel and killed whilst flying with the second, it is "*No bird;*" but if it is missed under the same conditions with one or both shots, the pigeon is lost. In short, if the shooter waits until the pigeon rises and is afterwards shot at, it is scored *to the shooter.* In the last case only may the gun be brought to the shoulder, but it is well understood that to be scored "killed" no bird must be killed except on the wing.
16. The shooter has a right to another bird if his gun miss-fires or refuses to go off through any fault not his own.
17. The pigeon is lost if the shooter has neglected to cock his gun, to load it, or to place on the cap.
18. If the first barrel misses fire, and the shooter fires the second, he loses his right to another pigeon, unless the second barrel also miss-fires.
19. If the second barrel misses fire, the shooter having fired and missed the bird with the first, he may claim another bird; but in that case both barrels must be loaded, the first with powder only, and neither barrel must be discharged until after the trap is sprung.
20. It is forbidden to shoot both barrels at the same time.
21. When the shooting is at single birds, and more than one bird is sprung, the shooter may refrain from shooting, and declare it no bird, but if he shoots the shot must in all instances be scored.
22. When shooting at double rises and more than two pigeons rise together, the same right is accorded as in Rule 21.
23. The pigeon to be scored must fall and be gathered within the boundary; if it falls without or within after having been outside the boundary, it is scored lost.
24. The boundary within which the birds must fall and be gathered to be scored good is shown by barriers or tape.
25. A pigeon to be scored good must be killed upon the wing, unless the second barrel is used. One person only must retrieve the bird, and must not employ any instrument to effect it.

26. All pigeons declared doubtful, that is to say, those which may be supposed to have sufficient strength to fly, although wounded, must, if one of the competitors demand it, be gathered immediately to be scored "killed."

27. Every pigeon that, after being shot, shall perch or settle on any tree, post, bar in the enclosure, or on the boundary rail, is scored lost unless it falls dead within the enclosure before the next shooter fires.

28. The fallen pigeon is not scored until it is gathered and brought home. The pigeons falling into the pool are scored good unless they rise and fly from the enclosure before the next shooter has fired his first barrel.

29. The shooter who leaves the mark after firing the first barrel loses the right of using his second.

30. All pigeons shot at behind the diagonal line of banners, whether killed or missed, are scored as lost to the shooter.

31. Each trap bears a number, and it is entirely dependent upon chance which trap is used.

32. In the handicaps, matches, and other shooting in which the bore and the load are specified, every shooter having used a gun of a different gauge or a different charge is excluded from that match, and loses his entrance-fee. And every shooter convicted of having shot at a less distance than the one fixed for him loses his right to the stakes or prize.

33. The standard gauge from which all the distances are calculated is the 12-bore. 11-bores go back half a mètre; the 10 bore one mètre. On the other hand, 14-bores advance half a mètre, and the 16-bores one mètre. Any gauge larger than ten is excluded, and no further advantage is allowed to any gun of less than 16-bore.

34. From 10 a.m. to noon, and from 1 p.m. to 2 p.m., the shooting is free. Each person shoots in turn, the distances and the gauges not being fixed. From 2 p.m. to 6 p.m. the shooting is exclusively reserved to sweepstakes.

35. A commission of five per cent. is reserved on all sweepstakes and wagers, but in matches between two shooters only the five per cent. commission will be charged on half the amount of stakes only.

36. The extreme distance in shooting for stakes is fixed at 30 mètres (33 yards); but the referee may augment it in the case of acknowledged superior shooting.

37. In shooting for sweepstakes at several pigeons, every shooter who may arrive after the shooting has commenced may enter, provided always that he scores as bad his pigeon or pigeons in the rounds finished before his arrival. For stakes at one pigeon only, no one can enter after the first round is concluded.

38. The winner of a stake amounting to 10 louis, free from all commissions, his entrance-fee included, goes back two mètres; if the stake is less, one mètre. This distance is imposed for the day only.
39. The same shooter cannot be put back more than three mètres the same day, although he may gain several stakes.
40. If the winner of a stake amounting to 10 louis, his entrance-fee included, is already shooting at thirty mètres, he stays at that distance, and the other shooters advance two mètres; or, if the prize is less, one mètre. By this rule, however, the shooter may be put back to thirty-two mètres instead.
41. Each shooter has his name written in a book placed on the stand, and the distance varies according to the judgment of the referees, who modify them according to their observations. Their decisions are without appeal.
42. The days for shooting are fixed by the committee, and may be seen at the Cercle des Patineurs or at the principal clubs of Paris.

FINES.

1. 100 frs. (£4), for shooting at a passing pigeon or other bird.
2. 20 to 100 frs., for every shot fired behind the diagonal line of banners; same fine for shooting the second barrel after leaving the mark.
3. 20 frs., for carrying a gun about the enclosure, unless called upon to shoot.
4. 20 frs., if the shooter receives or replaces his gun loaded and cocked.
5. Muzzle-loaders must be given to the shooter at half cock, or the armourer fined 20 frs.
6. Breech-loaders to be loaded at the mark, with the barrels toward the traps, and when the trappers have returned to their places, the shooter or the armourer not conforming to this rule will be liable to a fine of 200 frs.

BOUNDARIES.

The Hurlingham Club boundary is 90 yards from the centre trap, measured in a straight line.

The Gun Club boundary is 65 yards in a straight line from the centre trap, and is a solid fence about nine feet in height.

The Monaco boundary is 17 mètres (about $18\frac{3}{4}$ yards) from the centre trap, and is semicircular. The fence is of open wire trellis, and about 3 feet 6 inches in height.

The usual boundary is 80 yards from the centre trap, measured in a straight line.

The Monaco boundary leads to snap shooting.

PIGEONS AND TRAPS.

The pigeon generally employed for trap purposes is known as the Blue Rock. The best variety, the Lincolnshire Tin Blue Rock, retain the wild nature of the common blue Coast Pigeon. They are fed in Lincolnshire by the farmers in winter time, who also raise cotes for them at a good distance from their other buildings, as the wilder the birds and the nearer the coast they are raised the stronger and more hardy they are. The true Tin Blue Rocks afford the best sport, and are much the hardest to kill, being small in the body, quick in flight at starting, tough in their nature, and game to the death, especially the hens.

The Coast Pigeon (Blue Rock).

Other Blue Rocks are bred in Oxfordshire and Yorkshire in large quantities, but are inferior to the Lincoln birds.

Many of the so-called Blue Rocks are also imported from Antwerp ; in fact, the greater portion of the pigeons used for trap shooting are brought over from that port, and sold here as Tin Rocks. Some years ago a number of Tin Rocks were exported to France and Belgium for breeding purposes, and their offspring is now imported ; the foreign climate has not improved them, as they possess little of the gameness peculiar to the English bird.

The real Rock is not always of the same marking ; there are light and speckled Rocks.

The next best bird to the Blue Rock is the English Skimmer, which is chiefly employed at the second-rate clubs; and then true Antwerp pigeons.

A set of pigeon traps and pulling apparatus will cost about £35. Particulars may be obtained of Mr. Brown, of Nunhead, London. A single trap for practice with home-reared pigeons may be purchased for £2. All spring traps, plunge traps, &c., are of no use, and should not be tolerated in a match of any importance, nor allowed in a club.

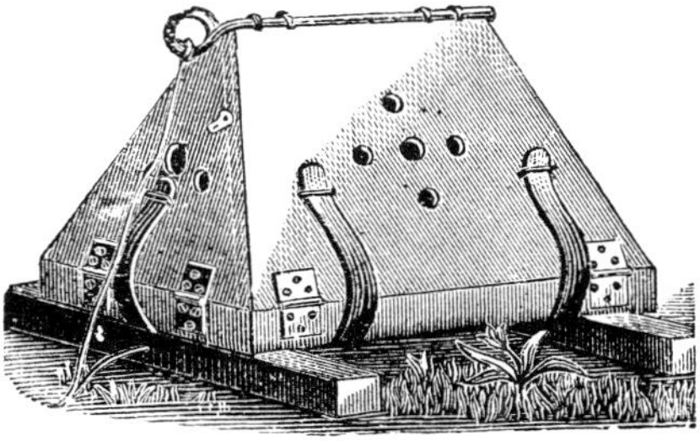

Leavenworth Trap.

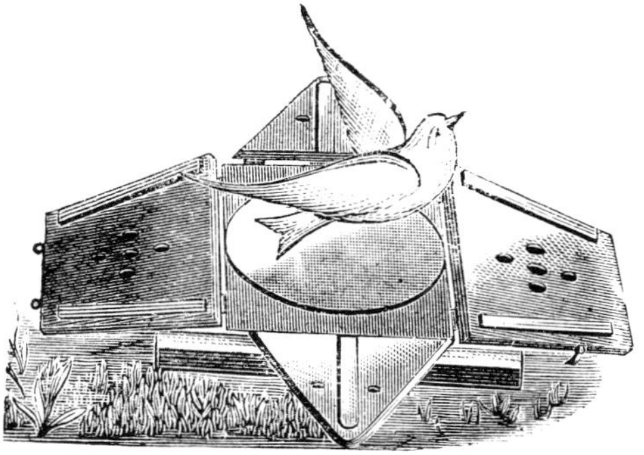

Leavenworth Trap, Open.

For private practice the "Leavenworth" trap, as illustrated here, is the best to force the pigeon to take wing at once, and it is a perfect trap for use with untrained pigeons. The author will be pleased to furnish particulars of this trap to any person desiring them, but no trap will make tame birds take flight like the wild, strong Rock pigeon, or as the trained pigeons of the best purveyors.

THE BEST RECORDS.

Undoubtedly the "Blue Ribbon" of the trap is the Triennial *Championnat Universel*, won by Mr. H. Cholmondeley-Pennell in 1886, and again by Mr. Blake in 1889. Both winners used Greener Guns.

The winners of the *Grand Prix du Casino* must also be considered amongst the best of trap shooters. This match is contested for by the best trap shots of all nations, and the birds are supplied by one of the most esteemed purveyors, whilst the Monaco boundary is acknowledged to be much in favour of the bird. The contest extending over several days also necessitates careful shooting over an extended period, and to kill thirteen consecutive birds without a miss, firing only at long intervals, is evidence of the ability of the marksman.

The following gentlemen have won the Grand Prix du Casino :—

Year. Winner of the Grand Prix.

1872—Mr. George L. Lorillard (American).
1873—Mr. J. Jee, V.C., C.B. (English).
1874—Sir Wm. Call, Bart. (English).
1875—Captain Aubrey Patton (English).
1876—Captain Aubrey Patton (English).
1877—Mr. W. Arundel Yeo (English).
1878—Mr. H. Cholmondeley-Pennell (English)
1879—Mr. E. R. G. Hopwood (English).
1880—Comte Michel Esterhazy (Hungarian).
1881—M. G. Camaueur (Belgian).
1882—Comte de St. Quentin (French).
1883—Mr. H. T. Roberts (English).
1884—Le Comte de Caspela (Italian).
1885—M. Leon de Dorlodot (Belgian).
1886—Signor Guidicini (Italian).
1887—Count Salina (Italian).
1888—Mr. C. Seaton (English).
1889—Mr. V. Dicks (English).
1890—Signor Guidicini (Italian).

Both in 1887 and 1888 the killing of twelve birds without a miss won the Grand Prix.

MATCH SHOOTING.

Perhaps the best score on record is that of Captain A. H. Bogardus, who on July 2nd, 1880, succeeded in scoring 99 birds out of the 100, the 47th bird falling dead out of bounds. This extraordinary score was made in a match with Mr. Rimmell, an English gentleman, for 250 dollars aside. Bogardus, 30; Rimmell, 28 yds.; 100 birds, 5 traps, weather fair, and birds in good condition.

The Captain used an English hammerless gun, and loaded with Dittmar powder in the first and orange lightning, No. 6, in the second barrel. Rimmell used the orange lightning powder, and both were restricted to $1\frac{1}{4}$ oz. shot.

The American birds are inferior to English Tin Blue Rock pigeons, and all pigeons are much stronger, wilder, and sharper in the early spring and winter than in summer.

Captain Bogardus, when in England, although shooting several matches, never approached the score just mentioned; the best performances being in a match with Mr. Dudley Ward, who shot a tie with him, each scoring 84 out of 100. Mr. Ward won in shooting off this match.

The match with Mr. Wallace, at the Gun Club Grounds, shot July 19th, 1878, resulted in a tie, each shooter scoring 69 birds out of 100. The following Wednesday the tie was shot off, resulting in a win for Mr. Wallace, he killing 71 birds to the Captain's 69. On July the 23rd, in the same year, the Captain shot a match with Mr. H. Cholmondeley-Pennell at the same grounds; the scores being—Captain Bogardus, 71; Mr. Cholmondeley-Pennell, 69. These scores are amongst the best ever made in England.

Dr. Carver made several matches with the best trap shots of England. He was beaten once by Mr. Heygate, of The Gun Club, in a match of 25 birds a side.

Dr. Carver tied with Mr. A. J. Stuart-Wortley in a match for £500 a side, shot at the Hendon Ground, December 8th, 1882—score, 83 each. This match was the more exciting from the fact that at the 50th bird the scores were equal, as they were again several times during the last part of the match and at the finish.

On February 7th, 1881, a match was shot off at the "Welsh Harp," Hendon, between Dr. Carver and Mr. W. Scott, a celebrated English wing shot. Dr. Carver used throughout the match a Greener Choke-bore Gun. Mr. Scott also used a Greener, but after the match commenced shot with the Greener and a London gun irregularly. The score was—Dr. Carver, 66 birds; Mr. Scott, 62. The stakes amounted to £400. The birds were the finest and quickest seen during the winter, and the weather was vile: the greater part of the match being shot in a blinding snowstorm, and a driving squall from the south-west.

In 1884 Dr. Carver and Captain Bogardus met and contested three matches. The first at Louisville, Ky., U.S.A. (100 birds, 30 yards rise, 80 yards boundary, Hurlingham Rules); score—Carver, 83; Bogardus, 82. The second at Chicago, Ill., U.S.A., same conditions as in the first match—Carver, 82; Bogardus, 79; at the 80th bird the scores were even, and remained so until the 90th, when Carver killed *all* his succeeding birds, and won a well-contested match by 3 birds. Third match at St. Louis, Mo., U.S.A. (50 double rises at 21 yards); score—Bogardus, 81; Carver, 79.

Some close, sharp shooting was exhibited in a match at the *Carteret Grounds*, Bergen Point, N.I., U.S.A., shot on the 19th of January, 1888, between Mr. C. Floyd Jones and Mr. Edgar C. Murphy. The match was shot under Hurlingham Rules, at 30 yards rise, 21 *yards* boundary. The wind was fresh and blowing across the grounds from the left, and the birds are said to have been good. Score—F. Jones, 78 ; Murphy, 77. Both shooters used light 12-bore English hammerless guns, and Schultze powder, and the time occupied was only 2 hours and 32 minutes. It is worth recording that although in matches where the two shooters shoot alternately 20 birds without a miss have been scored repeatedly, it is rarely that in one handicap, or several largely attended handicaps, any shooter will score this number without a miss. Dr. Carver's string of 50 birds, killed straight off, which he accomplished at Lynchburg, Va., U.S.A., with a Greener $7\frac{1}{4}$ lb. 12-bore gun, is his best on record in this line.

A final contest for the Championship of England Cup took place at Hendon on July 3rd, 1888, and resulted in a win for Captain Brewer, who killed 24 out of 25 birds, at 30 yards rise, and having thrice consecutively gained the prize against all comers, claimed the trophy as his own. Captain Brewer used a Greener gun in all contests.

In the contest for the American Field Champion Wing Shot Cup, 1890, Mr. Elliott, the holder, successfully defended it with a Greener gun, scoring 59 out of 60, 48 out of 50, and 94 out of 100 birds.

In Melbourne the Grand National Handicap was won with nine birds grassed in succession. Both the winner and *proxime accessit* used Greener guns ; as did also the winners of the Bathurst Champion Handicap, the Melbourne Challenge Cup, and the Melbourne Cup, the chief events of the Australasian pigeon-shooting tournaments.

SECTION II.

INANIMATE TARGETS.

As a pastime, the shooting at glass balls or bottles has long been practised in this country, but was developed and made a fashionable amusement in the United States by Mr. Ira Payne, Captain Bogardus, Dr. Carver, and other professional shots.

The inanimate targets now in use may be divided into two distinct classes—balls and "pigeons."

The balls, at first plain hollow spheres of colourless glass, were afterwards made of blue or amber glass, and filled with feathers; later the spheres were chequered to prevent the shot from glancing, and this stage of development is the highest reached by the glass ball. Balls made of various resinous compositions have been tried, and have a certain sale, but as there is difficulty in getting them sufficiently brittle they have not generally supplanted the glass balls. Other plans have been tried, as bell balls, puff balls, explosive balls, &c., but they have not proved successes commercially.

The traps to throw the balls have been wonderfully developed. From the modified catapult used at English fairs they have advanced to a

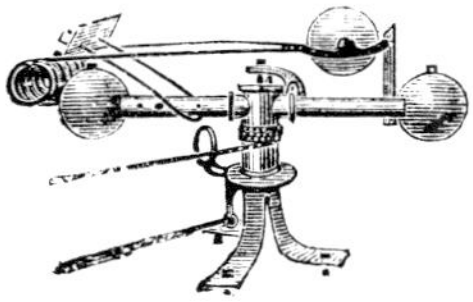

The "Carver" Revolving Trap.

rotating trap, which simply defies trickery on the part of the trap-puller or his assistant. The "Hatch" Trap was one of the first popular traps, then followed the Bogardus (illustrated here), the Carver, with a coil spring, instead of the flat coach spring of the Bogardus pattern, and then several revolving traps, of which the "Mole" and the "Card" are perhaps the best known.

The "Card" rotating trap will throw in any direction except towards the shooter, and is a simple and efficient machine.

The flight of a glass or other ball being inferior to that of a pigeon, and the traps failing to throw them so that they cannot be easily smashed

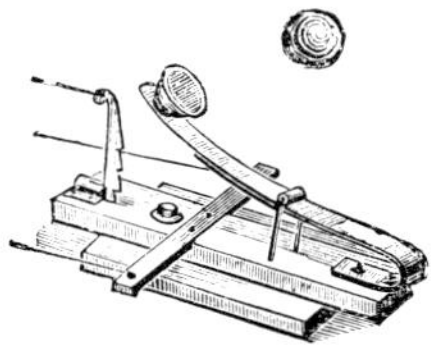

The "Bogardus" Trap.

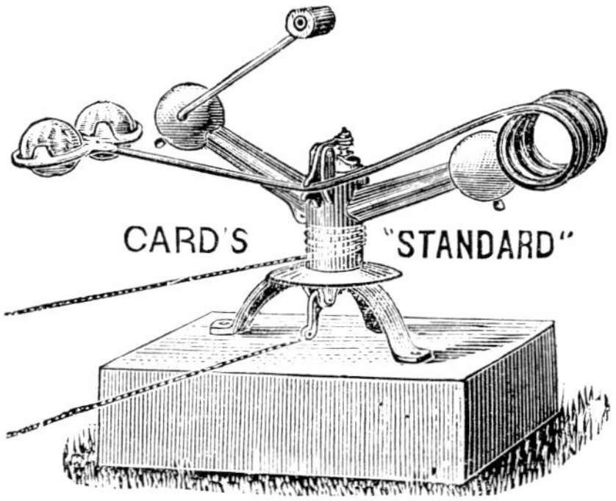

The "Card" Trap.

by a fair shot at all ordinary ranges, flat skimming targets were introduced to afford better practice. This form of inanimate target is said to have originated from the "Bussey Gyro," which was introduced about 1866, and made but little progress in popularity.

The Ligousky Clay Pigeon, which is now well known, is a moulded clay saucer, to which a paper tongue is attached, and the saucer is held by this tongue in the clamp of a specially constructed trap. When the trap is sprung the pigeon may be sent, almost horizontally, skimming through the air at a fair speed. This trap has had a large success, and its prominent features will at once be learned from a glance at the adjoined illustration.

The broken pieces of burnt clay are as indestructible as pieces of glass, and have been objected to by country gentlemen on that account, but the single or double brass pigeons made by Messrs. Kynoch & Co., of Witton, may be used with this trap, and as they fly better than the clay pigeons, and are easily gathered, their use may be recommended.

These pigeons are made upon the same principle as the puff balls already mentioned; they are more symmetrical than the clay pigeons, being stamped out of sheet brass. The base is of cardboard, and is firmly fixed by turning over the brass rim. The pigeons are filled with fine charcoal, and emit a cloud of dust when struck, and if fairly hit this is easily discernible. The double pigeons have not such a rapid flight as the single ones, and are made wholly of brass.

Bran puff balls made in two hemispheres, on the same principle are also to be obtained of Messrs. Kynoch, and they are more economical to use for practice than glass or composite balls.

The "Blackbird" variety of targets is composed of such mixtures as pitch and ashes, or rosin and plaster, and the materials are worked into shape in a semi-liquid form, in much the same manner as potters fashion clay on the wheel. Some of these birds are objectionable from the dull black colour, and as the packing required is excessive, heavy bills for carriage and numerous breakages during transportation have been the rule.

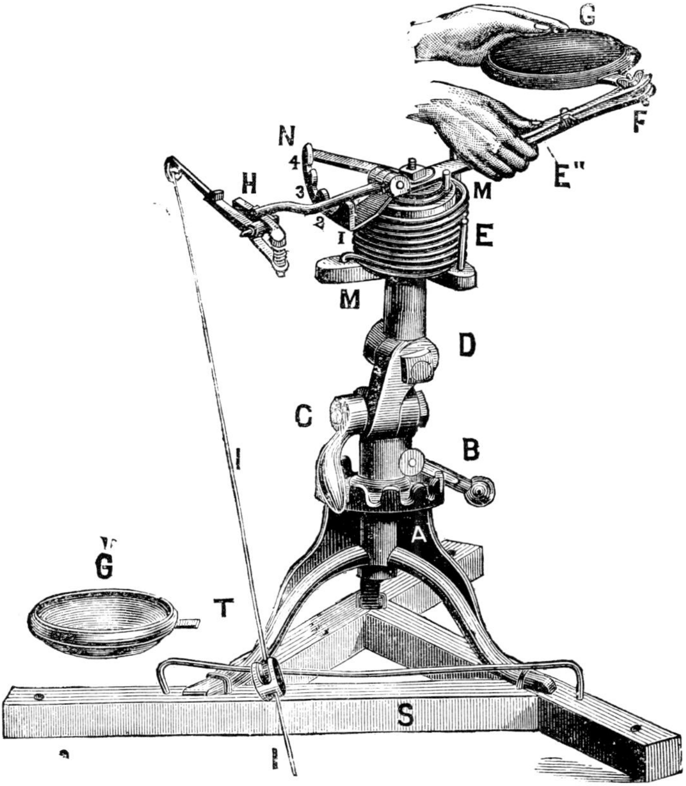

The Ligousky Clay Pigeon Trap.

The "Peoria" Blackbird and trap are illustrated here. It will be seen that the "bird," instead of having a tongue cemented to it, has two ears projecting from the solid body of the target, and these projections are not likely to be broken off in transit, or by careless trapping.

The "Niagara" Blackbird, the "Acme" (feather-filled) pigeons, the "Standard" targets, and others are of this type.

The "Peoria" Blackbird and Trap.

The "Blue Rock" Pigeon differs from most targets of the above description, inasmuch as it is made by pressure, the top is connected to the sides by a film-like connection, and this permits of the targets being so strongly made that breakages during transit are reduced to a minimum, whilst the target is shattered if struck by one shot. The target, of a deep black, has a yellow crown, making it readily discernible against any background.

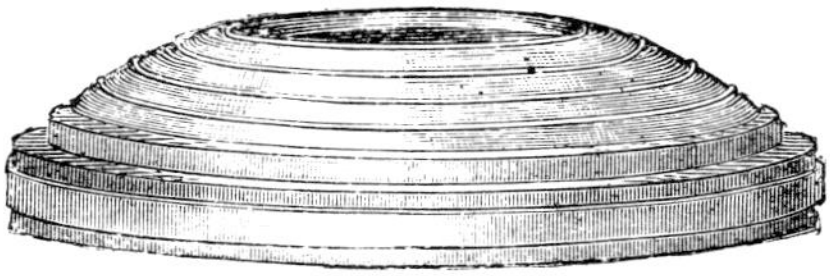

The Blue Rock Pigeon.

The trap used for the Blue Rock Pigeon is so constructed that it imparts to the thrown target a high rotary motion, and at the same time throws it with little friction. The trap also so throws the target that it has a steady flight and a good velocity, which is not perceptibly greater at the commencement than at the end of the flight.

The American traps are usually made with interchangeable parts, and are carefully fitted.

There are certain accessories necessary, such as cords, planks, and holding-down pins, and clubs will also provide themselves with pulling

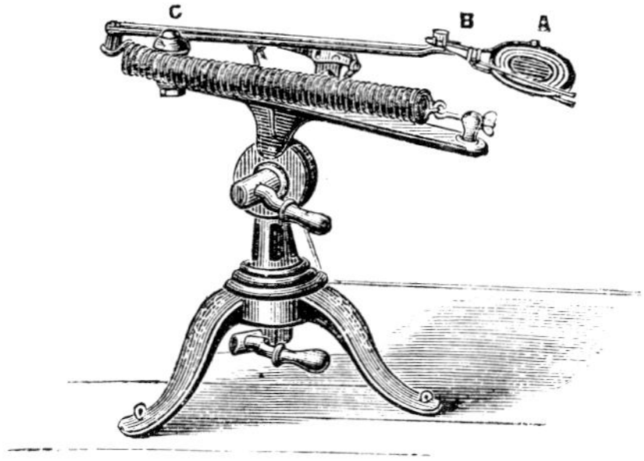

The Blue Rock Trap.

gear; and whether using three or five traps, will decide upon a definite method of determining in which order the traps shall be sprung. The use of dice or a trap-pulling indicator will effectually prevent collusion between the shooter and puller.

The following rules are those generally observed in inanimate target shooting, and with little or no alteration may be applied to matches with any number of traps:—

RULES FOR INANIMATE TARGET SHOOTING.

1.—DECISION OF JUDGES.—Two judges and a referee shall be appointed to judge all matches. If the judges cannot agree, the referee shall decide, and his decision shall be final.

2.—SPECIAL DUTIES OF REFEREE.—The referee shall see that the traps are properly set at the beginning of a match, and that they are kept in order to the finish.

3.—FLAGS FOR JUDGES.—Each judge shall be provided with a red flag and a white flag. They shall raise the red flag to indicate a "broken" bird, and the white flag to indicate "lost" bird; they shall raise both flags to indicate "no bird" or an imperfect bird; they shall also announce the score in a loud voice.

4.—KEEPING THE SCORE.—It shall be optional with the judges and referee to keep the score themselves, or appoint some one for that purpose, and the score thus kept shall be the official score, provided, however, that the referee must testify to the correctness of the score or scores made under his supervision if required.

5.—SCORE WITH INK.—All scoring shall be done with ink or indelible pencil; the scoring of a lost bird shall be indicated by an "0;" of a dead or broken bird by an "1."

6.—TRAPS.—All matches shall be shot from three traps set level, in the segment of a circle five yards apart. The radius of this circle shall be eighteen yards. The traps shall be numbered from No. 1 on the left to No. 3 on the right consecutively. All traps must throw the birds a distance of not less than forty yards.

Each trap must be tested for this standard distance before the shooting begins, and if any trap be found too weak to throw the required distance, a new trap or spring must be substituted.

7.—ADJUSTING TRAPS.—The lever or projecting arm of the trap shall be so adjusted that the elevation of the bird in its flight at a distance of ten yards from the trap shall not be more than eight feet nor less than four feet, and the angles of flight shall be as follows:—

No. 1 trap shall be set to throw a left quartering bird.
No. 2 trap shall be set to throw a straight away bird.
No. 3 trap shall be set to throw a right quartering bird.

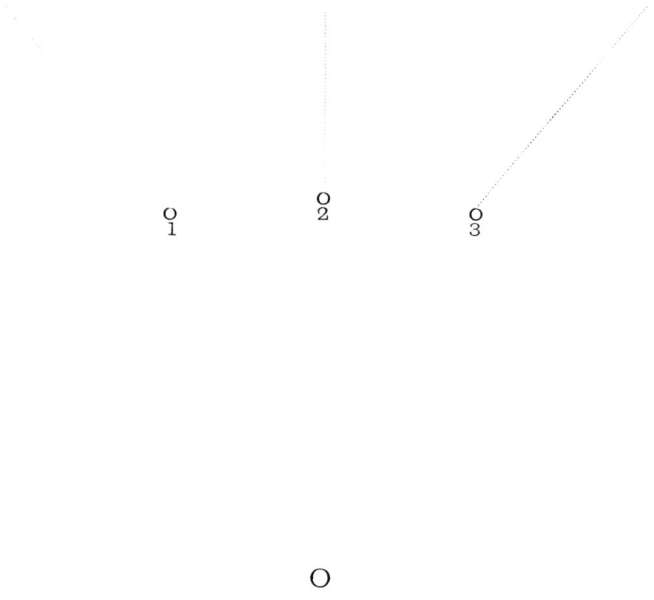

Diagram showing Position of Traps, and Directions for Throwing Birds.

After the traps are set for the above angles, if the bird for any reason shall take a different angle, it shall be considered a fair bird, provided the trap has not been changed.

8.—Pulling of Traps.—The trap-puller shall stand at least six feet behind the shooter, and when the shooter calls "Pull," the trap or traps shall be instantly sprung. In single bird shooting he should pull the traps indiscriminately, and not one, two, and three consecutively. He shall pull equally and regularly for all shooters. If the bird is sprung before, or at any noticeable interval after the shooter calls "Pull," he can accept the bird or not, but if he shoots the result shall be scored.

9.—Screens.—No screens shall be used. Back-stops may be provided for trappers, not to exceed ten yards from the end traps, and not to exceed three feet in height.

10.—Distance.—In single bird shooting the rise shall be 18 yards for 10-bore guns, 16 yards for 12-bore guns. In double bird shooting the rise shall be 15 yards for 10-bore guns, 14 yards for 12-bore guns.

All distances mentioned in these rules must be accurate measurement.

11.—Gun.—No gun of larger calibre than 10-bore shall be used.

12.—Loading Guns.— In single bird shooting only one barrel shall be loaded at a time, and the cartridge shall not be placed in the barrel until after the shooter has taken his position at the score; in double bird shooting both barrels to be loaded at the score.

13.—Position of Gun.—The butt of the gun shall be held below the armpit until the shooter calls "Pull." If this rule be violated, and the bird is missed, it shall be scored as a "lost bird." If it be broken, it shall be declared "no bird," and another shall be shot at.

14.—Broken Birds.—A bird to be scored broken must have a perceptible piece broken from it while in the air. The decision of the judges and referee on this point shall be final. No bird shall be retrieved to be examined for shot marks.

If a bird be broken by the trap, the shooter may claim another bird, but if he shoots, the result must be scored.

15.—Single Bird Shooting.—Each contestant shall shoot at three birds before leaving the score. If two birds are sprung at the same time, it shall be declared "no bird."

16.—Double Bird Shooting.—Both traps must be pulled simultaneously, and each contestant shall shoot at three pairs consecutively, thrown as follows :—

First pair from No. 1 and 2 traps. Second pair from No. 2 and 3 traps, and third pair from No. 1 and 3 traps. If only one bird is thrown it shall be declared "no birds," and if the gun miss fire on either bird it shall be declared "no birds." In each and all such cases another pair of birds must be shot at.

17.—TIE SHOOTING.—All ties shall be shot-off at the original distance and at the number of birds agreed on by the contestants. If, however, the contestants cannot agree promptly on this point, the referee shall fix the number, and his decision shall be final. The rules prescribed for single and double bird shooting shall prevail in tie shooting.

18.—CHALLENGE.—No challenge shall be considered, unless the parties challenging are contestants.

THE BEST RECORDS.

The best records made at inanimate targets are very much higher than anything obtained from live bird shooting. According to a list recently published there are more than a dozen shooters in the United States who have broken 100 of the inanimate targets without a miss, and the score made and recorded at a public competition. Upwards of fifty shooters have scored more than 90 out of 100.

"Young Nimrod," an English boy, has, with his 28-bore gun and $\frac{3}{4}$ oz. of shot, frequently scored 88-100 at clay pigeons.

In a series of twenty-five matches, at 100 clay pigeons each at each match, between Dr. Carver and Capt. Bogardus, 2,227 were broken by Dr. Carver, and 2,103 by Capt. Bogardus at 18 yards' rise. Dr. Carver made two scores of 100 each without a miss, and won nineteen matches, tied in three, and lost three. His lowest score was the first—72; and twenty of his scores exceeded 90 broken. Capt. Bogardus once scored 99, his highest, and three times 63, his lowest in this series of matches.

At glass balls still less skill is required; but the best record is Mr. Scott's—700 smashed consecutively with a Greener Gun. Dr. Carver, in a match with Mr. Scott, broke 9,737 out of 9,950 shot at; Mr. Scott 9,735 out of the same number. Out of the last 950 in this match Dr. Carver missed two only, and Mr. Scott three.

The quickest time recorded for breaking 100 glass balls with a shot gun is just under five minutes. Capt. H. Bogardus, the great American wing shot, made a match against time in December, 1879, and succeeded in breaking 5,500 glass balls in a few seconds less than 7 hours 20 minutes. The misses numbered 356. The Captain used an English gun with two pairs of barrels — one pair (10-bore) shooting 4 drams of powder and $1\frac{1}{2}$ oz. of No. 8 shot; the 12-bore pair were loaded with $3\frac{1}{2}$ drams and 1 oz. of No. 8 shot. During the match the Captain loaded for himself, and changed the barrels no less than fifty-five times. Three miss-fires only occurred in the whole series of 5,855 shots. The balls were all sprung from spring traps.

HINTS ON TRAP SHOOTING.

Trap shooting cannot be recommended as a profession. However good a shot a sportsman may be he will find so many uncertainties in trap shooting that it is doubtful if any person shooting continually will make trap shooting pay expenses.

At an ordinary bird, shot at under Hurlingham rules by an average good shot, the chances are five to two in favour of the shooter. To be considered a good shot, the number of kills must average more than 70 per cent. Mr. "Grace," at one time considered a reliable shot, with a Greener Gun once scored a percentage of 84·3 kills in a series of International contests. Other shooters have occasionally made a higher percentage in a short series of matches.

In a series of International matches, out of 1,120 birds shot at by thirty-six different shooters, 79·9 per cent. were killed, and this is about the average in matches between first-rate shots.

The following hints may be of use to young shooters who wish to try their skill in trap shooting:—Commence at a short distance—say 18 yards—at live birds; stand in an easy position, gripping the gun well forward with the left hand. This is a great aid in quick shooting. Do not stare at the trap which you think will give the most difficult shot to you, and if you do not particularly regard any trap so much the better. Do not say "Pull" until you are quite ready to shoot, and have your attention concentrated upon what you are about to do. When the bird gets up, up with the gun quickly but steadily, and immediately you have it in place at the shoulder it should be aligned at the bird, and the trigger pulled. Pigeon guns are so constructed that at 40 yards' range they will throw the body of the charge a few inches higher than the line of aim, consequently, at any distance up to 45 yards you will have the advantage over a bird rising in flight.

When shooting at 30 yards' rise this quality of the gun will be the more requisite, as to be a sure trap shot you will require generally to kill your pigeon within four or five yards of the trap, and for that distance the pigeon generally rises, and if he does not do so immediately will in all probability do so long before he is out of range. The happy medium between snapping just over the trap and "poking" after the pigeon must be sought.

In choosing a gun all will depend upon the rules under which it will be used, but it may be said that as a rule a gun of 7½ lbs. will be the thing. Let it be taken from the rack just before going to the mark, and let a point be made of loading and cocking it methodically. Quite a large number of birds are scored lost every year because the shooter has forgotten to cock his gun, move the safety off, or some other cause equally easy to prevent.

It is best to take no heed either of bystanders or trappers when going to the mark, and if one can be quite deaf to the shouts from the "ring" the score is likely to benefit.

In contending in a handicap it is the time spent in waiting between the rounds that tires and tries nerve and patience. At Monte Carlo a man may have to fire but nine times, and possibly have the whole of two afternoons in which to do it. Very much, therefore, will depend upon the temperament of the shooter. In contesting a match at 100 birds it must be remembered that the task will be trying to endurance; and if a light gun can be found which suits as well as a heavy one, the use of it will enhance the shooter's chance of success. The shortest time occupied by the match will be two hours and a half, and it may drag along for double that time.

In match shooting the percentage of birds killed will be greater than in handicap shooting, and unless the shooter knows by experience or former practice that he can kill on the average 90 birds out of 100, he will do best not to contest a match with the best shots of the day.

Drive straight to the shooting-ground, so as to arrive at the time the shooting is advertised to commence. Waste no time in plating your gun. If the results of the shooting at the target should not please you, you will lose confidence in your gun and gain nothing. You should ascertain that the gun shoots well, and that the cartridges are suitably loaded *before* you get to the shooting-ground. Keep yourself to the matter in hand, and pay no attention either to the remarks of other contestants or the "betting." Having won or lost, leave the ground at once. Unless the ground is one not often visited, trial shots before the serious shooting commences are not to be recommended. Upon visiting a town for the purpose of contesting the International Tournaments it is best to lodge at some distance from the shooting-ground, and to go there only so often as the business of the contests may require. Nothing is gained by constantly hanging about in the vicinity of the shooting-ground, nor by experimenting upon it. You should go to the ground to kill every bird at which you shoot; you must practise and experiment upon a private ground elsewhere.

A little experience will soon prove to the young shooter whether it is best to take or leave a pigeon which does not rise immediately the trap is sprung, and in other ways how to *win*, providing he is a good shot, and can keep in perfect health.

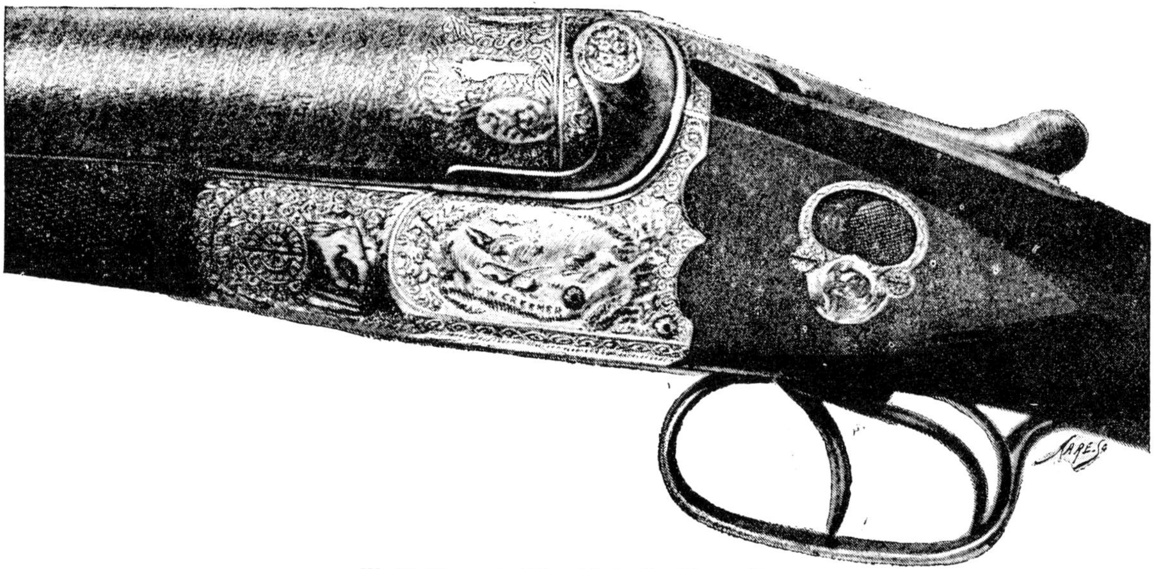

W. W. Greener's "Royal" Quality Ejector Gun.

APPENDIX.

STERLING STEEL.

STERLING STEEL AND HOMOGENEOUS METAL FOR GUN BARRELS.

Apart from the question as to which metal is *the* best for shot-gun barrels, it is possible to consider the merits of steel and iron alloys as material suitable for manufacture into gun barrels.

Steel of good quality has been proved to possess sufficient strength to resist the strain to which a shot-gun barrel is ordinarily subjected, and Messrs. Whitworth, by careful manufacture, have turned out barrels of a uniform quality of excellence.

The question of steel as a suitable metal for gun barrels has occupied the attention of gunmakers for several years, and after many experiments, a superior description of steel possessing valuable qualities has been adopted. "Sterling Steel" is the name given to it by the author.

The mode of manufacturing is not to draw the steel at all, as generally understood, but to forge out the barrel into a solid rod, and afterwards drill the whole length. Barrels so made are of close metal, stronger and denser than any obtainable by other means.

The "Sterling Steel" recommended is made of a homogeneous metal, of very fine quality, and admirably adapted by its great tenacity, or tensile strength, for use in gun barrels. It has been thoroughly tested by the author, as well as at the Government Proof House, with very heavy charges, viz., 28 drachms of powder and $4\frac{1}{2}$ ounces of shot, this charge being equal to seven ordinary charges of powder and four charges of shot. This test and many others it withstands perfectly.

"Sterling Steel" barrels are of sufficient strength for all practical purposes, and only in appearance are they at a disadvantage when compared with twisted Damascus and laminated steel barrels.

Unlike "cast-steel" barrels of the old type, "Sterling Steel" barrels bulge instead of breaking, and increased strain produces an open burst similar to that of a welded barrel instead of a sharp break or a longitudinal rip as is found to result with imperfect steel barrels. The quality of the metal is such that it will stand successively more than double the strain to which a sportsman can submit his gun with fair usage. And it will not "rip" or "crack," however sharp may be the explosive used.

Good steel barrels have the following advantages over the Damascus barrels: they are less liable to *honeycomb* from the corrosive action of gunpowder, they are not so easily bent from rough usage as twisted barrels, and being harder are not so likely to be injured when accidentally dropped. Many guns have been ruined by this cause alone.

There being no welds in these barrels they are absolutely free from *greys* or flaws of any description. They may be relied upon as being equal to those manufactured by the Whitworth Company; and further, they do not add to the cost of the gun, whereas Whitworth barrels cost more, and "Sterling Steel" is always obtainable without the annoying delays which often result when a special barrel is required of Whitworth metal.

The bulges were caused by placing a small charge of shot between two felt wads (first a thick felt, then the shot, then the thin felt) at the spot where the bulges are, and firing an ordinary charge from the gun.

The burst was effected by increasing the charge of shot between the wads, the bulges appeared about fifteen inches from the breech end after firing.

As many as five thick wads may be placed in any part of the barrel, and the gun fired without causing a bulge, but we find by experiments that even the small quantity of $\frac{1}{8}$ oz. shot placed between wads at any place in the barrel, will cause a bulge even as near as nine inches from the breech. The different sizes of the bulges in the illustration were caused by different charges of shot. The shape of the burst indicates the extent of bulging before bursting. The illustration is reduced to half-size.

As to the shooting qualities of steel barrels, it is not claimed that they shoot better than those of ordinary gun iron, as good shooting is only obtained by skill in boring, but steel being harder, they should retain their shooting qualities longer than those of softer metal. Other steels may possess the tenacity required of the material for shot gun barrels, but none have yet proved so reliable. The alloys of iron, manganese steel, nickel steel, aluminium steel, and other materials possessing essential features are not yet made of such uniform quality as will admit of their adoption by gunmakers of reputation for making into shot-gun barrels, although undoubtedly of such an alloy the guns and rifles of the future will be manufactured.

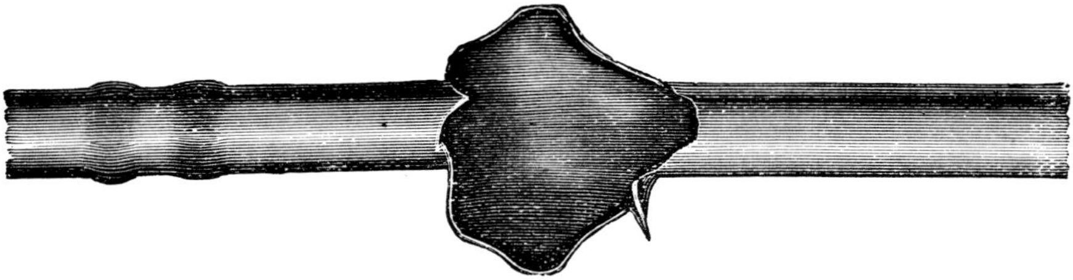

Bulges and burst in a Barrel of "Sterling Steel."

"THE 'GREENER' SHOT COUNTER."

Further particulars of the "'Greener' shot counter," mentioned at page 146, have been so frequently demanded that the adjoined illustration and description of this simple contrivance will doubtless be welcomed by those cartridge loaders to whom the machine is still unknown.

By drilling holes of the requisite size and depth in a sheet of hard brass, a species of trowel is formed, in which the pellets of shot will stick when the trowel is forced into a mass of shot and slowly withdrawn. For

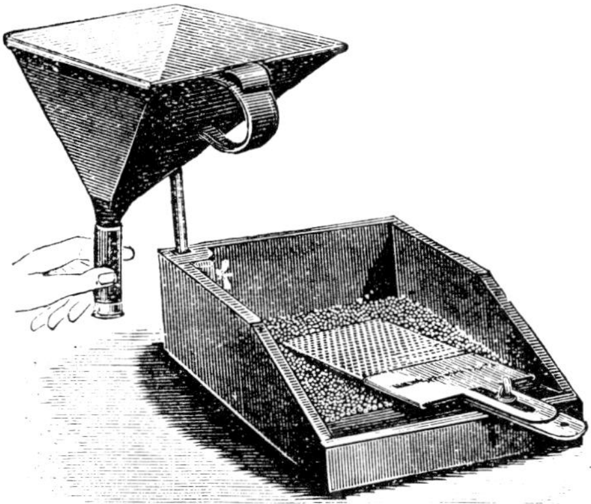

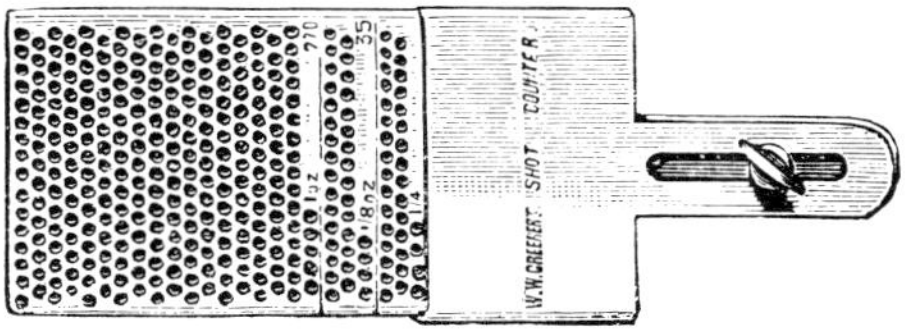

the use of sportsmen a pattern is now made, with a sliding cover, by which the number of holes exposed in the trowel may be varied, according to the charge of shot it is wished to load into the cartridge.

As already stated, the charge of shot is in this manner regulated by *number*—270 pellets of No. 6 weigh one ounce, and in good shot if this number should turn the scale, it will be found that 269 will not; 304 holes for 304 pellets are allowed for the $1\frac{1}{8}$ oz., or standard charge, and beyond this charge the graduation may be varied with the greatest nicety.

In order to obtain exact results it is absolutely necessary that the *same number* of pellets are loaded into the cartridges; if the number varies—although the weight of the charge may not—there will be a variation in the pattern, which will probably prove misleading, and as it is easily avoidable by using the counting trowel, the weighing of charges of shot should be discontinued.

The use of the trowel with a plentiful supply of shot, and the specially made square, or ribbed, tundish, enables any one to load cartridges accurately with great speed and ease, and it is a method far in advance of any system of weighing or measuring shot charges. A glance at the trowel as it is withdrawn filled from the shot-box will show whether any undersized or misshapen pellet is on the trowel, any irregularity in size, shape, or number of the pellets being instantly discernible, as each pellet occupies a separate cell and is seen distinctly.

The price, with box, tundish, and stand, is so trifling—viz., 21s.—that every one who loads cartridges should give this method of loading a trial. It will be found satisfactory, and a properly made trowel will load millions of cartridges before the holes become so worn that it has to be discarded.

THE GIFFARD GUN.

This weapon, about which so much has of late been said and written, is a French invention. M. Giffard, whose chemical and mechanical experiments have always been regarded with interest, has long sought to utilise carbonic acid gas as a force for the propulsion of rifle projectiles. Having succeeded in liquefying the gas, and being fully conversant with its properties of expansion, he substituted a reservoir of liquefied gas for the receiver of the old and well known air-gun. The fall of the hammer upon pulling the trigger of the lock, liberates a small quantity of the liquefied gas, which, when in contact with the air, expands to its fullest extent with the force of an explosive, and propels the bullet, pellets, or whatever projectile may be placed in the barrel. When exhausted, the reservoir is detached from the gun and replaced by another.

The reservoirs must be filled at the factory, and the gas specially prepared.

At present the invention has not passed the experimental stage, and it would be rash to predict either that the Giffard system will supersede firearms, or that the sporting-gun has nothing to fear from its new competitor.

INDEX.

INDEX TO ILLUSTRATIONS.

PRICES OF

W. W. GREENER'S

Breech-Loading Guns, with all the Latest Improvements.

No.		£	s.	d.
200.	—The "DOMINION GUN," Top-lever, Double-bolted, Snap Breech Action, with extended rib, Laminated or "Siemens'" Steel Barrels, either choked on W. W. Greener's world-renowned method, or improved cylinder, to order. In this Gun all the value is put into the barrels, locks, and shooting	9	9	0
	Illustrated on page 10.			
207.	—Made with double grip, with lever over-guard as illustrated on page 7	9	9	0
201.	—The "TRAP" Gun, similar to the Dominion, but with a third grip in extension rib, and neatly engraved ...	10	10	0
	The shooting of each Gun guaranteed.			
203.	—A Top-lever Gun, with either Bar or Back action, Laminated or "Siemens'" Steel Barrels. As illustration on p. 9	10	10	0
202.	—Superior make and finish	12	12	0
108.	—Greener's "FORESTER" Gun, Treble-wedge-fast Cross-bolt, Rebounding locks, front action engine-turned rib, neatly finished and slightly engraved	13	13	0
109.	—The "FANCY FORESTER," superior finish and well engraved	14	14	0

Good shooting guaranteed.

The illustration on page 12 shows the style, but the engraving is not so elaborate on the Forester.

No.		£	s.	d.
107.—	The "Far Killer." This popular and celebrated Treble-wedge-fast Gun is now fitted with either Laminated Steel or Siemens' Steel Barrels, tastefully finished and engraved	16	16	0
106.—	Superior make and finish	21	0	0
105.—	Better qualities	26	5	0
104.—	,, ...	31	10	0
103.—	,, ...	35	15	0
102.—	,, ...	42	0	0
101.—	,, ...	47	5	0
100.—	Royal Quality, finished similar to the Gun illustrated on page facing 195	57	15	0

SINGLE-BARRELLED GUNS.

	£	s.	d.
Single-Barrelled Breech-loading, good, sound, serviceable, double-grip, under-lever Gun, plainly finished	5	5	0
Ditto, with engraving	6	6	0
Top-lever, Snap Breech mechanism, neatly finished, without engraving	6	6	0
Ditto, engraved	7	7	0

Better Qualities, £8 8s., £9 9s., and £10 10s.

All 10-bore Guns £1 1s. extra for size.

8-bore Duck Guns, as illustrated on page 95; weight, 10½ to 11 lbs. for 2 oz. shot, 11½ to 12 lbs. for 2¼ oz. shot; to order.

Prices from £15 15s.

With Treble-wedge-fast Action, from £21.

		£	s.	d.
4-bores, weighing 15 to 20 lbs., Double Grip	from	£21	0	0
Ditto Ditto Treble-wedge-fast ...	,,	£26	5	0

HAMMERLESS DUCK GUNS.

		£	s.	d.
8-bore Single Treble-wedge-fast, as illustrated on p. 96...	from	£26	5	0
4-bore ,, ,, ,, ...	,,	£31	10	0
8-bore Double-barrelled ,, ,, ...	,,	£31	10	0

Length of Barrel—8-bores, 32 to 36 inches; 4-bores, 36 to 42 inches.

Gun Cases, with or without Implements, are supplied to order and charged extra.

W. W. GREENER'S HAMMERLESS GUNS.

The "Facile Princeps" lock mechanism, the celebrated Treble-wedge-fast Breech mechanism, Patent Side Safety Bolt, with which these Guns are fitted, have given every satisfaction in all parts of the world. All these Guns are most carefully manufactured by W. W. Greener, and the utmost value is put into the barrels and lock and breech action mechanisms, so that W. W. Greener can, with confidence, recommend them as being safer, handier, and more durable than the Hammerless Guns upon other systems which are now offered at same low prices.

No.	£	s.	d.
28.—A plainly-finished, well-made Hammerless Gun, without engraving, English, Damascus, or Siemens' Steel Barrels	16	16	0
27.—Better quality, neatly finished and engraved	21	0	0
26.—With Greener's Sterling Steel or English Laminated Steel Barrels	26	0	0
This type of Gun, 12-bore, can be made with 27-inch barrels as light as $5\frac{3}{4}$ lbs.			
25.—Ditto, superior quality and finish	31	10	0
24.—Ditto, better quality, tastefully finished and engraved ...	36	5	0
23.—Best quality, with hand-welded English Laminated Steel, Sterling Steel, or Damascus Barrels	42	0	0
22.—Ditto, finest quality, elaborate finish, and engraved with artistic designs	47	5	0
21.—Royal Quality, finished as illustration facing p. 195 ...	57	15	0

All Hammerless Guns are fitted with W. W. Greener's Side Safety as shown on pp. 20 and 23, usually independent, but can be made automatic, to order, £1 1s. extra. Top Safety as illustration on p. 17, made to order, £1 1s. extra.

W. W. GREENER'S SELF-ACTING EJECTOR GUNS.

No.	£	s.	d.
5.—The highest development of the Sporting Shot Gun, fitted with English Laminated Steel or Sterling Steel Barrels, neatly engraved and well finished, as illustration on pp. 23 and 24	36	15	0
4.—Better quality and finish	42	0	0

No.		£	s.	d.
3.	Best quality, engraved and finished similar to illustration on p. 23 ...	47	5	0
2.	Ditto, more elaborately engraved and finished ...	52	10	0
1.	Royal Quality, being the very best, both as to material and workmanship, the barrels Greener's Sterling Steel, Best Laminated Steel, or finest English Damascus, and the Gun most tastefully finished and artistically engraved with sporting scenes (as illustration facing p. 195), or fine scroll design ...	63	0	0

EXTRA BARRELS.

Extra Barrels usually cost half the original price of the Gun, but no barrels can be properly fitted for less than £10 10s.

New Barrels to old Guns on same terms. It is necessary *for us* to have the old Gun to fit them to.

LIGHT GUNS AND SHORT BARRELS.

Guns with 30-inch barrels as light as $6\frac{1}{4}$ lbs. made to order.
„ 28 „ „ 6 lbs. or just under, made to order.

(Guaranteed to shoot nearly as well as a 30-inch.)

A Gun with 27-inch barrels, weighing $5\frac{3}{4}$ lbs., is the lightest and shortest-barrelled Gun we recommend for ordinary shooting. Shorter Barrels and lighter Guns made to order.

PRICES FOR REPAIRS, &c.

All repairs are executed in the best manner possible, by skilled workmen conversant with the mechanism of all kinds of arms in all stages.

Re-stocking—	Best Hammerless Gun, with Safety Bolt on side or top ...	6 gs.
„	Second Quality „ „ „ ...	5 gs.
„	Guns 30 to 40-guinea quality ...	5 gs.
„	Guns 20 to 25-guinea quality ...	4 gs.
„	Guns 15-guinea quality ...	$2\frac{1}{2}$ gs.
„	"Keeper's" Guns and Muzzle-loaders from ...	21s.

(The above prices include Blueing the Furniture and Browning the Barrels.)

Price List of Repairs.

New Strikers and Springs... ... ...per set		5s.
Converting Pin to Central Fire ... ...3 gs. to		5 gs.

(According to the value of the work done.)

Converting Side Safety to Tumbler-bolting Automatic ... 1 g. &		2 gs.	
Converting Muzzle-loaders to Breech-loaders ... ...7 gs. to		12 gs.	
	£	s.	d.
Cleaning Locks, &c., of Gun	0	5	0
Stocks Altered in Bend ... ... 10s. to	1	0	0
Stocks Lengthened by Fitting Horn Heel Plates	0	10	0
Re-browning Barrels, Polishing Chambers and Muzzles ...	0	10	0
Re-boring Barrels, Taking out Rust, and Testing Shooting £1 to	1	10	0
Improving Shooting of Old Barrels	1	10	0
Taking out Dents from Barrels ... ... 10s. to	0	15	0
Taking Choke out of Barrel, Shooting and Adjusting same, each Barrel	0	12	6
Facing or Jointing-up Shaky Actions ... ... 10s. to	1	5	0
Thoroughly doing up and Cleaning Gun, Re-chequering, Blueing, Browning Barrels, &c. &c. ... ... each, £1 to	2	0	0
Cleaning only	0	5	0
Blueing only	0	5	0
New or extra Mainsprings ... ...per pair, 5s. to	0	8	0
,, Top-Lever Springs...	0	2	6
New Cocks or Tumblers for Hammerless Guns each, 5s. to	0	10	0
Crests, Initials, or Monograms Engraved on Thumbpiece each, 5s. to	1	0	0
Gun Cases initialed or named, from ... ... each, 3s. to	0	6	0

Other Alterations or Repairs at Lowest Rates.

ALL PRICES FOR REPAIRS ARE NET AND FOR CASH.

Sportsmen not living in the neighbourhood of London should send their Repairs direct to the Factory at Birmingham.

The Novelty of the Year.

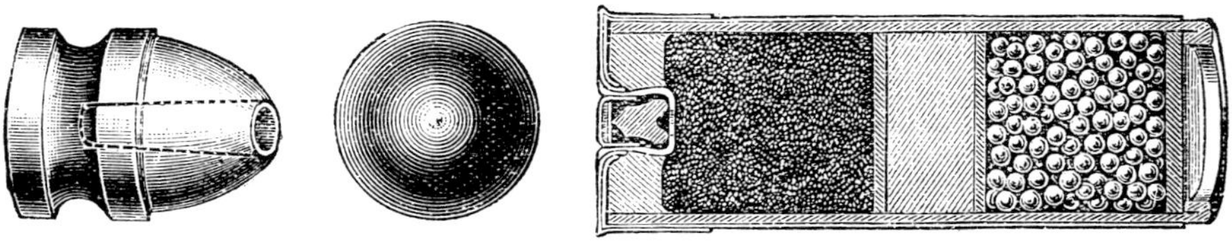

W. W. GREENER'S PATENT CHOKE-BORED RIFLE.

A PERFECT SHOT-GUN AND A PERFECT RIFLE IN ONE.

Performs equally well with conical and spherical ball, solid and hollow bullets, light and heavy charges of powder.

Shoots shot of any size, making regular patterns from the open shooting of the cylinder to the dense pattern of the choke.

The grooving is invisible, cannot foul, and is as easily kept bright and clean—*throughout its whole length*—as is the barrel of the shot-gun.

This weapon is distinctly different in *principle* to the rifled choke-bores offered under fancy names, and it achieves what none other has yet accomplished.

It has the accuracy and force of the heavy rifle combined with the lightness and handiness of the shot-gun. It is, without a doubt, the best weapon for snap-shots at large game, and if used as a shot-gun when loaded with expansive conical bullets, it is the best weapon for deer-drives, boar-drives, also tiger, bear, and elk shooting, or whenever advantage has to be taken of a snap-shot at driven or moving large game.

12-Bores as light as 7 lbs.

A fine assortment of 12-Bore Hammerless Choke-Bored Rifles at prices from 25 guineas. Prices for other bores 28 to 4 cal. on application.

W. W. Greener's "Sterling Steel," which rifles perfectly, and is stronger than Damascus, is used for all barrels of the Choke-Bored Rifles.

W. W. GREENER'S RIFLES.

The Cape Rifle and Shot-gun combined, the right barrel being 12 or 16 bore, and the left ·450-bore for long ranges, sighted 700 to 1,000 yards. Made in same quality as Guns, but charged 3 guineas extra in all grades.

DOUBLE EXPRESS RIFLES.

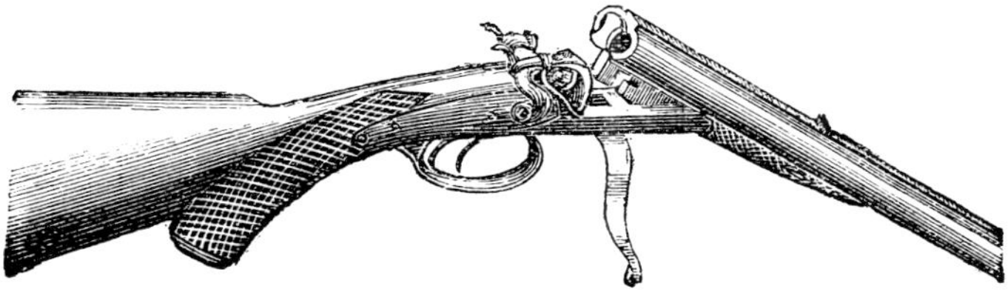

THE "JUNGLE" RIFLE.

This weapon is made, as shown in the accompanying illustration, with the well-known Under-Lever Double-Grip Action to either of the usual Express Cartridges of ·400, ·450, or ·500 bore.

The Rifle is guaranteed Sound and Accurate, and is a Marvel of Cheapness. All W. W. Greener's Rifles are shot and sighted accurately to 100, 150, and 200 yards.

	£	s.	d.
Price, without Case	**17**	**17**	**0**
,, for Case complete *from*	3	3	0
Extra for 577 Express	2	2	0

Higher Grades in proportion.

THE "FORESTER" RIFLE.

This Rifle is made like the "Forester" Gun, with the W. W. Greener's "Cross-bolt" Action, which makes a perfectly safe Rifle for any charge of Powder, and is a great boon to those who are used to a Top-Lever, which is much quicker and handier than the Double-Grip Under-Lever.

Made in same quality as the "Jungle" Rifle, and accuracy guaranteed.

	£	s.	d.
Price for ·400, ·450, ·500 bore ...	**21**	**0**	**0**
Extra for 577 Express	2	2	0
Case complete *from*	3	3	0

Finer Grade Rifles with the Treble-wedge-fast "Cross-bolt" action, **30**, **40**, and **50 Guineas.** Cases extra.

W. W. GREENER'S HAMMERLESS EXPRESS RIFLES.

These Rifles are fitted with W. W. Greener's celebrated Treble-wedge-fast Cross-bolt Breech Action and his well-known "Facile Princeps" lock mechanism. The Rifles are all carefully built, and are guaranteed safe under all possible conditions. Each weapon is carefully sighted and adjusted, and the accuracy guaranteed; private range for testing—500 yards. The mechanism is of the simplest and strongest, the accuracy unapproachable; and these weapons may be implicitly relied upon by sportsmen.

EXPRESS RIFLES.

Bores ...	·360	·400	·450	·500	·577	
Weights ...	7	8	9	10	12	lbs.

·450 and ·500 bores, with 28-inch barrels, are kept in stock of these weights. Lighter or heavier rifles built to order.

No.	Prices for Rifles.	£	s.	d.
414.	—Well-made Hammerless Express Rifle, no engraving, plain finish, steel barrels, fitted with Intercepting Safety Bolts and all modern improvements	31	10	0
413.	—Best quality Hammerless Express Rifle, plainly finished. Steel or Damascus barrels	36	15	0
412.	—Best quality Hammerless Express Rifle, neatly engraved and finished	47	5	0

This Rifle is also made to order in 16, 12, or 10 bore for conical and spherical bullets.

411.	—Best quality Hammerless Express Rifle...	57	15	0

This Rifle is also made to order in 16, 12, or 10 bore for conical and spherical bullets.

Hammerless Self-Acting Ejector Rifles, in best style throughout, fitted with Greener's fine Sterling Steel or Damascus barrels ... ... 45 gs. & 60 gs.

Cases and Implements from £5 5s.

W. W. GREENER'S ELEPHANT RIFLES.

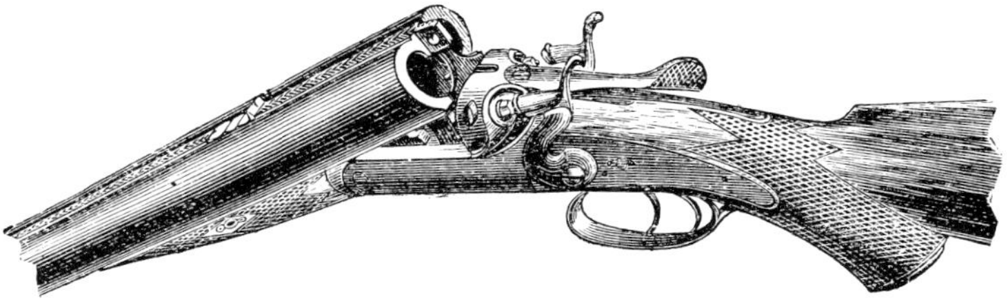

DOUBLE ELEPHANT RIFLE.

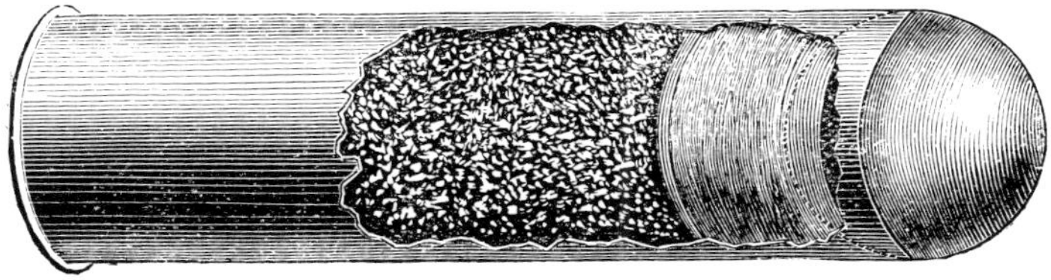

8-BORE BRASS CARTRIDGE FOR ELEPHANT RIFLE.

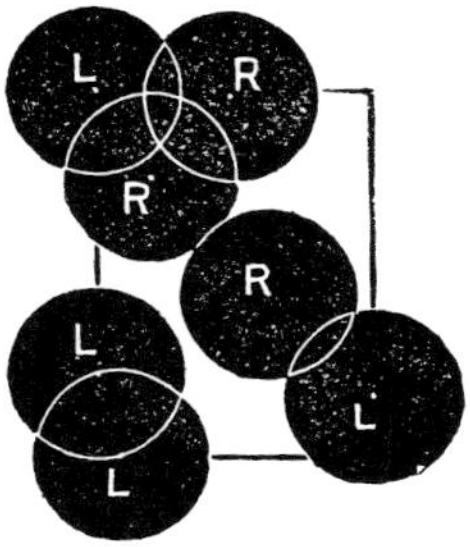

Fac-simile Diagram of Shooting made at Helsingfors, by the Editor of *Sportens*, with a W. W. Greener Treble-wedge-fast Rifle ·450-bore; 110 grains powder; hollow bullet, 270 grains; muzzle velocity, 12,700 feet per second. Seven consecutive shots at 100 yards. Actual size of square, 1 by $\frac{11}{16}$ inch.

A Diagram never equalled by any Sporting Rifle.

LARGE-BORE RIFLES FOR ELEPHANT SHOOTING.

There is but one action fit for Rifles of this calibre; it is the "W. W. Greener Treble-wedge-fast." This is what the author of "Thirteen Years amongst the Wild Beasts of India" wrote of his Large-bore Rifles:—

"Greener's Treble-wedge-fast Rifle, 8-bore, double-barrelled, weighing 17 lbs., and Greener's Treble-wedge-fast, double smooth-bore, 4-gauge Ball Gun, weighing 19 lbs., were built for me by W. W. Greener in 1874, since which time they have been in continual use, and have been fired several hundreds of times with 12 drams and a 2-oz. bullet, and with 16 drams and a 4-oz. bullet, respectively. They have never required any repair, and are still sound, and the breech-actions as close as when they left the factory. They have given me every satisfaction.

"S. P. Sanderson,
"Superintendent of Government Elephant
"*London*, 1884." "Keddahs, Dacca.

These Guns have since been used by Mr. Sanderson with great success, and are now in splendid condition. No sportsman living has had more experience than Mr. Sanderson of Indian elephant hunting, and the wear to which his rifles have been subjected is greater than is likely to fall to the large-bore rifles of any Indian or African sportsman. W. W. Greener's Large-bore Rifles and Ball Guns have also been used with great success by many well-known hunters in Africa, India, and East India, and reference is permitted.

Prices of Large-bore Elephant Rifles.

No.	£	s.	d.
427.—The "Dacca" Elephant Gun. Treble-wedge-fast 4-bore Double-barrelled Ball Gun, of best quality throughout, and as recommended by Mr. Sanderson, plainly finished	52	10	0

This weapon may also be had rifled for spherical bullets instead of heavy smooth bore, at same price.

425.—The "Keddah" Elephant Rifle, 8-bore, double-barrelled, as supplied to Mr. Sanderson and many noted sportsmen, sighted for *50 yards*, 100 and 150 yards, 16 to 17 lbs. weight, plainly finished, but of best quality throughout	47	5	0
426.—The "Tonquin" Rifle, 8-bore, double-barrelled, 24-inch, rifled for spherical and conical bullets and explosive shells, sighted and tested to 150 yards, of best quality and lightest weight, $12\frac{1}{2}$ lbs. to 13 lbs., anti-recoil heelplate, rebounding locks, steel barrels, as built for the most experienced hunter in the French colonies	42	0	0

Also are built to special order, Single-barrelled Breechloading 4-bore and 8-bore Rifles, or Ball Guns, prices of which will be furnished on application. Solid brass cases are recommended, and are always used unless otherwise specified on the order.

MARTINI SINGLE EXPRESS & LONG-RANGE RIFLES.

Of Single-Barrel Rifles now in the market, the Martini holds the first position, and deservedly so. For Simplicity of Construction, Strength, Solidity, and perfect Safety it is unequalled. The Rifle has everlasting wear, and is thoroughly reliable ; it is, too, a Hammerless, ejecting the fired case, thus embodying the latest inventions without extra cost.

The Martini is a Rifle which can always be relied upon, the Lock being unaffected by any climate, and the Stock never being known to break ; it is, moreover, the Cheapest Rifle made that will stand heavy Express or other charges.

Prices of Express Rifles for ·360, ·400, ·450, ·500 bore—

	Plain Quality.			Medium.			Best.		
	£	*s.*	*d.*	£	*s.*	*d.*	£	*s.*	*d.*
With Safety Bolt, Straight Hand	**6**	**6**	**0**	**7**	**7**	**0**	**9**	**9**	**0**
,, ,, Pistol Hand...	**7**	**7**	**0**	**8**	**8**	**0**	**10**	**10**	**0**
577 Express	**8**	**8**	**0**	...			**10**	**10**	**0**

Martinis, not Government marked, for explorers, hunting expeditions, ships' chests, etc., at special low prices for quantities.

Carbines, same price.

These Rifles are also largely made to use the British Government ammunition at above prices, and serve either as a Long-Range or Express Rifle. They are in great demand in South Africa, where they are the general favourites.

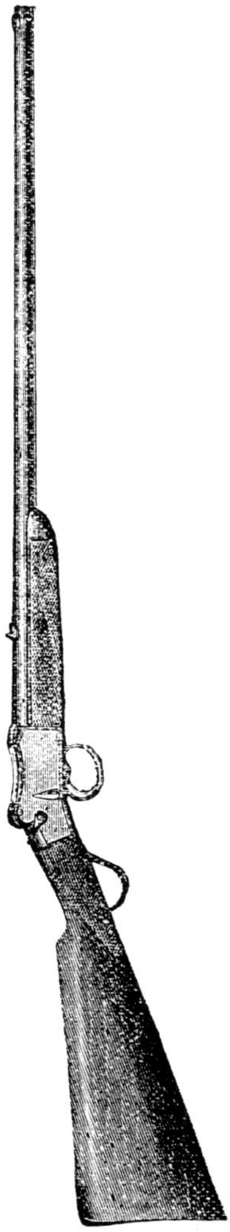

Volunteer, Match, and other Long-Range Rifles,

Shot up to 500 yards, at **5, 6,** and **7 Guineas.**

A large stock of all sizes kept, so that they may be supplied without delay.

All kinds of Volunteer Requisites and Shooting Accessories supplied.

THE NEW REGULATION ·303 REPEATING RIFLE.

W. W. Greener is licensee under the Lee-Speed Patents to manufacture Magazine Rifles identical with those in use by Her Majesty's forces. These rifles are of first-class workmanship, accurately sighted, and shot and regulated at W. W. Greener's private range (500 yards).

By an arrangement with a contractor to the Honourable Board of Ordnance, ammunition of the same make and quality as issued by the Ordnance Department can be supplied with these weapons.

It is believed that the high velocity, low trajectory, and enormous penetrative force possessed by this rifle when used with the Regulation ammunition, will lead to its adoption for sporting purposes generally.

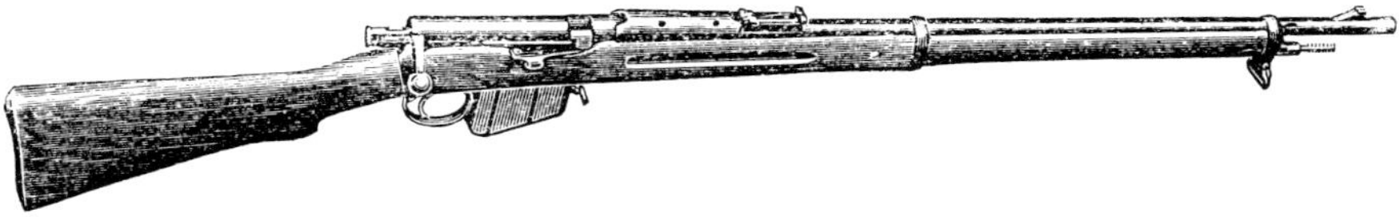

No.	£	s.	d.
·303 Repeating Rifle with one detachable magazine, regulation Infantry pattern, price	10	10	0
The same rifle with sporting stock and chequered fore-end, price	12	12	0
·303 Repeating Rifle, Artillery pattern, price	10	10	0

Ammunition 20s. per hundred, loaded with the Regulation charge, 85-grain compressed pellet, and 215-grain steel-covered composite bullet, in solid-drawn brass cases with hermetically-sealed pinions. Smokeless powder cartridges can now be supplied.

THE NEW ·303 MARTINI.

REGULATION PATTERN.

This rifle is now being manufactured for Her Majesty's Government. The breech mechanism is that of the Martini-Henry Rifle adapted for the

New Small-Bore Barrel of the Magazine Rifle (Enfield-Lee), and the range and sighting are identical with the range and sighting of the new Service arm.

W. W. Greener is now supplying new rifles upon this principle, each rifle shot-regulated and guaranteed to possess the same shooting qualities as the rifles made in Government factories.

Price £8 8s. per Rifle.

Superior-finished, Sporting style, £10 10s.

Steel or brass cleaning-rods, 2s. 6d. each.

W. W. GREENER,

Arms Manufacturer.

Contractor to the British and Foreign Governments.

RIFLE SHOOTING EXTRAORDINARY.

THE NEW MAGAZINE RIFLE IN KIMBERLEY.

Extract from "The Kimberley Independent."

"Under the auspices of the genial Mr. A. P. Walshe, a trial was made yesterday for the first time in Kimberley of the famous New British Government Magazine Rifle, which has provoked so much discussion in European military circles since its tentative adoption by the War Office. The Rifle, viewed from the practical colonial standpoint, seems to fulfil every condition except weight. A curious thing connected with the gun is that it has scarcely any recoil, and but a slight report. The penetration is very great, and yesterday two shots fired at a heavy stinkwood post, by Mr. A. J. Wright, went right through, and left a hole as clean and unsplintered as if it had been made with a gimlet. At 200 and 300 yards the point-blank shooting was all that could be desired. At 400 yards Mr. Wright made 5 consecutive bulls; at 500 yards he made 3 bulls and 1 centre; at 800 yards, 2 centres and 1 bull. Time being short, the remaining trials were made without a marker; but at 1,000, 1,500, and

1,800 yards, Mr. Wright and Mr. Finlason succeeded in hitting the target 5 times out of 6 shots. In order to try the extreme ranges, 4 shots were fired at 2,500 yards, and Mr. Wright missed the target by only a couple of yards, while Mr. Finlason highly astonished himself by dropping a bullet just over the target on to the mound. Half in jest, the party went back to Mr. Wright's house, and aimed at the Diamond Fields Horse cannon target; and with the aid of an opera-glass the bullets, on two occasions, were distinctly seen to strike over the target. The distance is about 3,500 yards. This is an extraordinary range for any rifle. Exactly the same results were obtained from the Martini Rifle, which has been adapted to the Magazine ·303 Rifle Cartridge. The Rifles were made by W. W. Greener, who is licensee under the Lee-Speed Patents to manufacture them."

ROOK RIFLES

With Martini actions, 4, 5, and 6 guineas; with Top-lever and Side locks, very superior quality, 8 to 10 guineas.

REVOLVERS.

Greener's Bull-Dog Revolver.

·320, ·360, ·380, and ·450 bores, £2 2s. each

Cheaper qualities, 30s., 21s., 12s. 6d.

Very superior Self-Extracting Army Revolver. Same as supplied to the British Government £6 6 0

CARTRIDGES FOR SHOT-GUNS.

To obtain good shooting, cartridges of the first quality are indispensable. To meet a growing demand, W. W. Greener has now arranged for the sale of his loaded cartridges at various Agencies in England and the Colonies.

All W. W. Greener's cartridges are loaded in one way, with perfected machinery, whilst the pellets of the charge are *counted* in by means of a simple machine of W. W. Greener's invention. Mistakes cannot occur, and the number contained will be found not to vary a single pellet, 304 being *counted* in as a charge of $1\frac{1}{8}$ oz. of No. 6 *chilled shot*, of 270 to the ounce. The same exactitude is observed in choosing and fitting the wadding, and "E.C.," "Schultze," "S.S.," and such powders are *weighed* or carefully measured.

The powder used is or one uniform quality—the very best procurable, and it will be found to give better shooting than that generally obtainable. Every batch of powder received is tested by W. W. Greener before being issued to sportsmen, and if above or below the requisite strength is rejected.

The prices are the lowest possible for the best quality cases and material, and careful loading. Each cartridge sold by W. W. Greener will be found perfect.

Prices of Loaded Cartridges at the Works, or at 68, Haymarket, London, and various Agencies.

The "Smokeless Cartridges," 12-bore, per 100.

"*Sporting Life*" Cases, "Schultze," "E.C.," or "S.S.," 42 grs. and $1\frac{1}{8}$ oz., 10s. 6d. cash, or 11s. 6d. booked.

Or in red or buff Cases, "Schultze," "E.C.," or "S.S.," 45 or 47 grs., and $1\frac{1}{8}$ oz., 11s. cash, or 12s. booked.

Ditto, 47 grs. and $1\frac{1}{4}$ oz., 11s. 6d. cash, or 12s. 6d. booked.

The "Prize" Cartridges, 12-bore, per 100.

Green or blue paper Cases, 3 drams & $1\frac{1}{8}$ oz., 9s. 6d. cash, or 10s. 6d. booked.
$3\frac{1}{4}$ drams & $1\frac{1}{8}$ oz., 10s. cash, or 11s. booked.
$3\frac{1}{4}$ drams & $1\frac{1}{4}$ oz., 11s. cash, or 12s. booked.

If in "Perfect," "Perfectly Gas-tight," "Grouse," or other fancy Cartridge Cases, the prices will be 6d. per 100 more; also $2\frac{3}{4}$-in. Cases 6d., and 3-in. Cases 1s. per 100 extra.

The "London Club" Cartridge.—In Eley's Green Cases, with special large Caps, for best Black Powder and Soft Shot, at 9s. per 100 for 12-gauge. For "Schultze" or "E.C." powders, in Eley's Special Cases, at 10s. per 100 for 12-gauge.

Small Bores.—From the above prices deduct 6d. per 100 for 16-bores, and 1s. per 100 for 20-bores, with usual charges; other sizes and loads at special prices.

The above prices are for delivery at 68, Haymarket, London, S.W., or at the Works, St. Mary's Square, Birmingham. No Railway Carriage or PACKING CASES are included in these prices.

Strong Packing Boxes, to hold 100, 200 300, 400, 500, or 1,000 Cartridges.
3d. 6d. 9d. 1s. 1s. 3d. or 2s. 6d.

Delivery, Carriage forward or paid, to any Railway Station at home or abroad, by arrangement.

Empty Cartridge Cases.—Ordinary Best quality, Blue, Red, Green, and "*Sporting Life*" Cases, or Eley's, Joyce's, or Kynoch's, at 40s. per 1,000. Fancy Paper Cases and "Perfects" from 42s. 6d. per 1,000. "E.B." Brown Cases, 30s. per 1,000. 20-bores, 1s. per 1,000 less.

Gunpowder, Shot, and Wadding.—Very best Black Gunpowder, No. 4 Alliance, retail at 2s. per lb.; "Gamekeepers'" Powder, 1s. 6d. per lb.; **Genuine Newcastle Chilled Shot, 8s. per 28 lbs. "Schultze," "E.C.,"** and "**S.S.**" at 8s. per lb. Felt Wadding, 3s. to 5s. per lb.; Pink-edged ditto, 5s. per 1,000; Waterproof and Card ditto, 1s. and 1s. 6d. per 1,000.

TERMS.—On no account will the Cash with Order Prices be allowed unless the Cash accompanies the order; otherwise the Quarterly Booked Prices will be charged.

INANIMATE TARGETS FOR MATCHES AND PRACTICE.

The "Blue Rock" Target is the best imitation of the living bird.

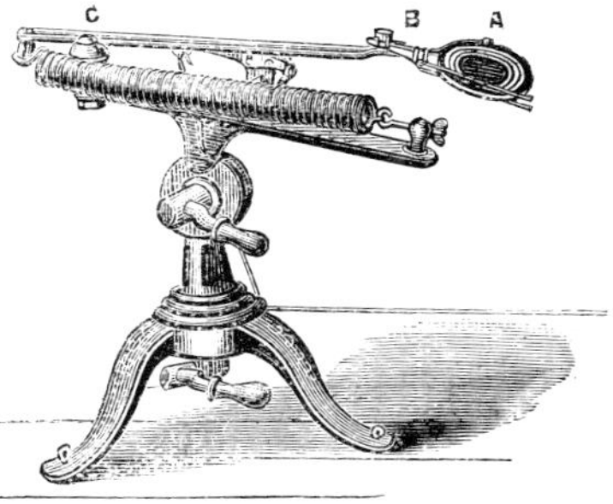

THE "BLUE ROCK" TARGET TRAP.

	Net Cash only.			
The Standard "Blue Rock" or other First-quality Trap ...	£2	2	0
,, "Blue Rock" or other First-quality Targets per 100	0	8	4
,, Cord for Trap	0	1	0

No allowance made for breakages in transport.

These Traps and Birds are obtainable only of W. W. Greener, at any of his addresses.

W. W. GREENER'S GLASS-BALL TRAPS.

THE "CARVER."

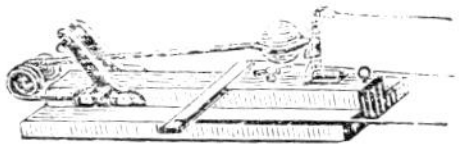

Fixed or Stationary Trap. Very strong and easily adjusted.

Price for single ball	£1	1	0
To throw two balls ... *extra*	0	5	0

THE ROTATING TRAP.

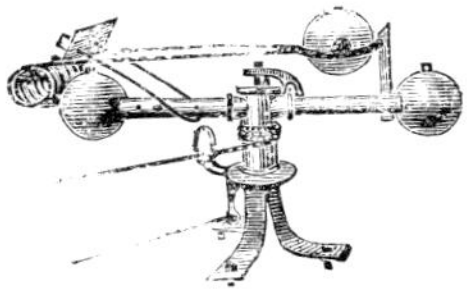

Throws the ball any direction but towards the shooter. Can be used as a Stationary Trap if desired.

Price for single	£2	2	0
If to throw two balls	2	7	0

Pigeon Traps, best quality, as used at the Gun Club, £2 2 0 each.

HONOURS TO THE GREENER GUN.

Winning Gun at the GREAT LONDON GUN TRIALS of 1875, beating 102 Guns by all the best makers of Great Britain and Ireland.

Winning Gun, BOGARDUS *v.* SOUTH, South killing 86—100; Philadelphia, 1876.

Winning Gun, CHOKE-BORES *v.* CYLINDERS. Silver Cup, £50; London, 1877.

Winning Gun, GREAT LONDON FIELD TRIAL OF EXPLOSIVES, 1878.

Winning Gun, GREAT INTERNATIONAL PIGEON CONTESTS, DIEPPE and BADEN-BADEN, 1878.

Winning Gun, GRAND PRIX DU CASINO, value £1,328; Monaco, 1878.

Winning Gun, GREAT LONDON GUN TRIALS OF 1879.

Winning Gun, GRAND PRIX DE MONTE CARLO; Monaco, 1879.

Winning Gun, SPECIAL FIRST ORDER OF MERIT; Sydney, 1879.

Winning Gun, CHAMPIONSHIP OF ENGLAND, 1880.

Winning Gun, Gold Medal, New Zealand Exhibition, 1882.

Winning Gun, GRAND PRIX DE LA MARCHE, 1880.

Winning Gun, CHAMPIONSHIP OF THE WORLD CHALLENGE CUP, 1881.

Winning Gun, Three Gold Medals awarded, Melbourne, Adelaide, Cleves, 1881.

Winning Gun, Highest Score ever obtained in a Public Pigeon Match, CARVER *v.* CRAWSHAY, 93—100 1881.

Winning Gun, Highest Scores ever obtained in a Public Glass-Ball Shoot; CARVER *v.* SCOTT.

Winning Gun, Gold Medal and Silver Cup; SOUTH AUSTRALIAN GUN CLUB, 1881.

Winner, CARVER *v.* BINGHAM, at 600 Pigeons, London, 1882.

Winning Gun at Hurlingham, 1882.

Winning Gun, Chicago, 1882.

Winning Gun, GRAND INTERNATIONAL CHAMPION CUP, 1882.

Winning Gun, Hendon, 1882.

Winning Gun, CARVER *v.* BOGARDUS, for $1,000; Chicago, 1883.

Winning Gun, CARVER *v.* BOGARDUS; 25 Matches at Clay Pigeons score, 2,227—2,500, United States, 1883.

Winning Gun, 5 Objets d'Art Reims, June 1883.

Winning Gun, Virgoe's Trophy Melbourne, 188[illegible]

Winning Gun, GRAND PRIX DE PARIS, 1884.

Boulogne, 1884.

First Award, Calcutta, 1884.

First Award and Gold Medal Nice, 1884.

Winning Gun, Cups and £49[illegible] LONDON INTERNATIONAL, 188[illegible]

Ballarat Annual, 1st, 3rd, & 5th, 1885.

Winning Gun, Highest Average, LONDON INTERNATIONAL, 1885.

Winning Gun, Highest Average, London, Season 1885.

Winning Gun, UNIVERSAL CHAMPIONSHIP (also 3rd and 4th), Monaco, 1886.

Winning Gun, Consolation Prize, Monaco, 1886.

BARCELONA, 1888.—GOLD MEDAL.

Valuable Prizes won with the Greener Gun.

PIGEON SHOOTING EXTRA-ORDINARY.

The Highest Shooting on record in a match between the two experts, Captain Brewer and Mr. J. A. R. Elliott, at Jersey City, March 26th, for 1,000 dols. a side. Captain Brewer scored 99 out of 100, the last bird falling dead out of bounds; Elliott killed 92.

Both shooters used Greener Guns.

AMERICAN FIELD CHAMPION WING-SHOT CUP.

CHAMPIONSHIP OF THE WORLD CHALLENGE CUP, Richmond, Va., 1887.

THE BELGIAN CHAMPIONSHIP, Brussels, 1887.

FIRST AWARD, Adelaide, 1887.

BATHURST CHAMPIONSHIP HANDICAP, 1887; 59 Competitors; Greener's Guns 1st, 2nd, and 3rd.

GOLD MEDAL and DIPLOMAS, ST. PETERSBURG GUN TRIALS, 1887.

CHAMPIONSHIP OF ENGLAND, 1888.
CHAMPIONSHIP OF THE WORLD CHALLENGE CUP, LONDON, 1888.
MELBOURNE 1st, 1888. UNIVERSAL CHAMPIONSHIP—"2nd Time"—1889.

THE GRAND PRIX DU CASINO.
1878.

THE GRAND PRIX DU CASINO, OBJET D'ART and £791; 97 Competitors, 1891.

THE PRIX D'OVERTURE,

THE PRIX DE MONTE CARLO, 1891.

AMERICAN FIELD CHAMPION WING-SHOT CUP, won five times successively, 1890-91, by Mr. J. A. R. ELLIOTT, with a GREENER Gun, in one match killing 98 birds out of 100.

THE GRAND PRIX DU CASINO,
1891.

THE GREAT LONDON FIELD TRIAL, 1875.

THE FIELD CUP.

At this important trial the superiority of Choke-bores over Cylinders was fully demonstrated, and their reputation thoroughly established. The principal class was for 12-gauge Choke-bores, for which the "Field" gave a Silver Cup, value 40 guineas.

In this class there were thirty-three competitors, with 68 guns. Each gun was fired fifty times at 40 yards, and twenty-five times at 60 yards. Our gun won, making an average pattern of 214, and penetration of 206·5 at 40 yards. In No. 1 Class we came out first with our 8-gauge Choke-bore, also first and second with our 10-bores. In Class No. 4 our little 20-bore, weighing $5\frac{1}{2}$ lbs., came out winner by several points. This trial was followed by a Wear and Tear, at which each gun was fired 2,500 times, at the rate of 60 shots per hour, ours coming out first in both pattern and penetration. Thus our guns won all the classes for Choke-boring.

In the Field Trial for Explosives, 1878, a full Choke-bore gun of our make made the most wonderful shooting on record, the average pattern of 25 shots being 220·8, and penetration 154·32, with $1\frac{1}{8}$ oz. No. 6 shot.

Again, in 1879 our guns carried off the prizes in the 12- and 20-bore classes.

CHOKE-BORES *v.* CYLINDERS, LONDON, 1877.

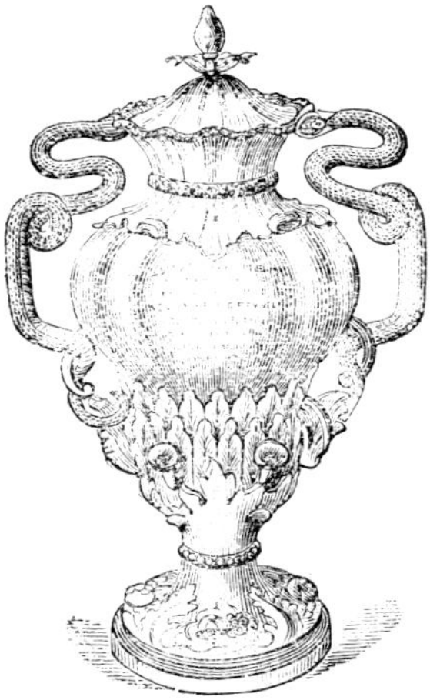

THE FIFTY-GUINEA CUP, PRESENTED BY MR. J. PURDEY.

This trial was shot off at the Gun Club, Notting Hill, May 23rd, 1877.

There were nine guns on each side, at five birds each at 30 yards, and five each at 40 yards.

The Choke-bores won the first day by four birds. The best score was made by Mr. H. C. Pennell, who killed 5 birds at 30 and 3 birds at 40 yards. It was noticed that Mr. Pennell with his Greener gun brought his birds down in splendid style at from 60 to 70 yards distance; he also used but $3\frac{1}{4}$ drs. powder, whilst $3\frac{1}{2}$ drs. was used by some of the other competitors. The prize for this match was a 50-guinea cup, given by Mr. Purdey. We had only one gun in this match, used by Mr. Pennell, against fourteen by the first London makers, and three by provincial makers. The next day a sweepstake was shot for, the same sides competing.

The Choke-bores were first by 14 birds. This trial fully proved that Choke-bores were the best weapons for this shooting, and they have since come to be generally used at all the gun clubs.

At the Gun Trials, 1886, Leavenworth, U.S.A., 50 guns, 45 yards, 24-inch circle, Greener 12-bore Gun won by 12 pellets, left barrel, and 13, right barrel, beating many 10-bores.

PLAN OF ROUTE TO W. W. GREENER'S WORKS.

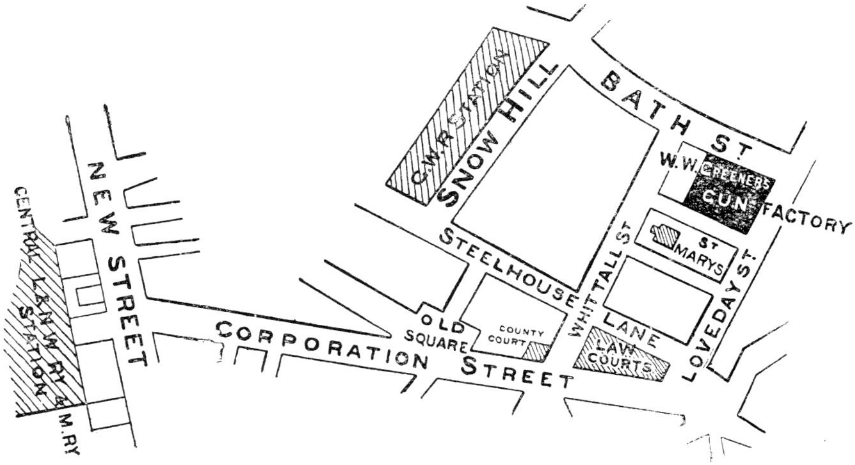

W. W. GREENER begs to inform his patrons and sportsmen that, to enable him to meet his largely increasing business, he has lately extended his gun manufactory, situated in Birmingham, fronting St. Mary's Square, Loveday Street, and Bath Street, where gunmaking in all its branches may be inspected. It is the largest and most complete in England, having an area of 37,000 square feet; also a 50-yard shooting range, try guns, and all modern improvements.

The largest stock in Great Britain of high-class and moderate price guns and rifles always on hand, suitable for Home and Foreign markets.

Gentlemen visiting Birmingham should leave the Central Station by the front entrance, Stephenson Place, walk straight up Corporation Street, and take the first turning on the left past the New Law Courts, Loveday Street (see plan), which is about ten minutes' walk.

Hours of business, 9 till 6; Saturdays, 9 till 1. Closed on Bank Holidays and day following.

W. W. GREENER'S GUN FACTORY, ST. MARY'S SQUARE, BIRMINGHAM.

W. W. GREENER'S ORDER FORM.

Name

Address

Description of Breech Action

No. and Page in Price List

Price and Terms

Gauge

Length of Barrels

Weight of Gun

Shooting—Left Barrel

,, Right Barrel

Stock Bend at Bump, at Comb

,, Length at Centre, at Toe, at Bump

,, Cast-off

If Pistol Hand, &c.

Fore-end , Locks

Finish , Sight

Engraving , Rib

Safety required

When to be finished

Gun Case No. , Price

Remarks and Shipping Instructions

[*Tear out this Form, and when* **each** *space is filled in, send to* W. W. GREENER.]

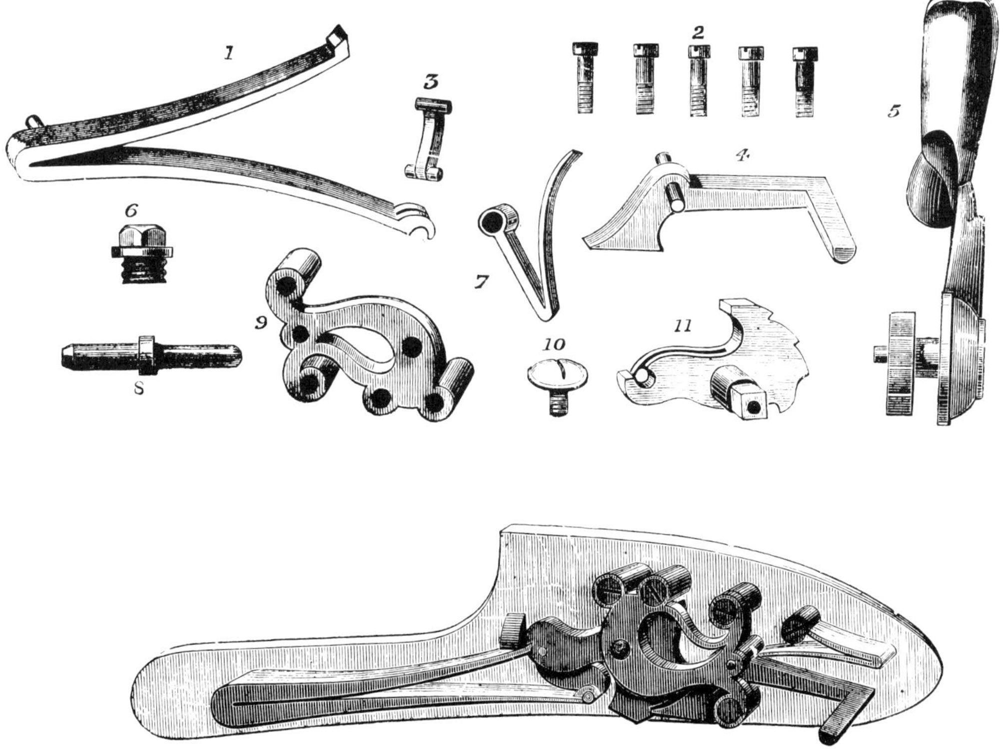

1
2
3
4
5
6
7
S
9
10
11